Agrometeorological Approach to Sustainable Agriculture

The Authors

Dr. R.S. Mishra obtained his M.Sc. (Ag.) Plant Pathology from Allahabad Agricultural Institute, Allahabad and Ph.D (Environmenal Botany) from Jamia Hamdard University, Delhi. Dr. Mishra is presently working as an Assistant Professor, Plant Pathology at N.D. University of Agriculture and Technology, Kumarganj, Faizabad (UP). He has more than 17 years of teaching and research experience. He obtained training from CRIDA, Hyderabad; IITM, Pune; and IARI, New Delhi on different aspects of Agrometeorology and Plant Disease Management.

Dr. Mishra also developed a protocol for Mushroom Cultivation and Management of Practices of *Coprinus* spp. in paddy mushroom. He has published more than 60 research papers in journals of repute.

Dr. Raj Bahadur obtained his Master degree in Plant Physiology from C.S. Azad University of Agriculture and Technology, Kanpur and Ph.D. in Crop Physiology from N.D. University of Agriculture and Technology, Kumarganj, Faizabad (U.P.). He worked at ICAR-Indian Institute of Pulses Research, Kanpur in Indo-Australlian collaborative research project. He is presently working as Scientist, Plant Physiology at N.D. University of Agriculture and Technology, Kumarganj, Faizabad. He is actively engaged in teaching and research programme. He has published more than eighteen research papers in journals of repute and thirty two articles in national magazine.

Agrometeorological Approach to Sustainable Agriculture

Dr. R.S. Mishra

Dr. Raj Bahadur

2016

Daya Publishing House®

A Division of

Astral International Pvt. Ltd.

New Delhi – 110 002

Cataloging in Publication Data--DK

Courtesy: D.K. Agencies (P) Ltd. <docinfo@dkagencies.com>

Mishra, R. S. (Assistant professor of plant pathology), **author.**
Agrometeorological approach to sustainable agriculture / Dr. R.S. Mishra, Dr. Raj Bahadur.

pages cm

Includes bibliographical references and index.

ISBN 978-93-5130-915-4 (International Edition)

1. Meteorology, Agricultural. 2. Sustainable agriculture.
I. Bahadur, Raj (Plant physiology scientist), author. II. Title.

S600.5.M57 2016   DDC 630.2515   23

Published by   : **Daya Publishing House®**
A Division of
Astral International Pvt. Ltd.
– ISO 9001:2008 Certified Company –
4760-61/23, Ansari Road, Darya Ganj
New Delhi-110 002
Ph. 011-43549197, 23278134
E-mail: info@astralint.com
Website: www.astralint.com

Laser Typesetting   : **Classic Computer Services**, Delhi - 110 035

Printed at   : **Sanat Printers**

Preface

In a country like India where the population graph has been continuously rising and has now touched the one billion mark, sustainable agriculture growth is the need of the hour to meet the ever-growing food requirements. The indiscriminate exploitation of natural resources has but only aggravated the problem as among other things it has caused serious ecological imbalances. Large-scale deforestation to bringing more and more area under cultivation has played havoc as it has brought about serious topographic and climatic changes. The conditions of soil erosion, irregular monsoon often culminating in wide-spread drought and flood, and environmental population which adversely affects both flora and fauna, are largely due to deforestation. Besides, the mismanagement of water resources, the inadequate and unbalanced use of fertilizers, inclement weather conditions, and the menace of pests and insects to standing crops, are the other major factors responsible for the poor and inadequate production of food.

It is, therefore, imperative to advise effective means to overcome the above natural and man-made problems which tend to jeopardize our food prospects beyond measure. It is not that we are oblivious of the above threats and in any way indifferent towards resolving the same as various measures are being adopted both at the government and private levels to combat them. But there is still a lot to in the above direction and a greater awareness has to be created among the masses to enable them to understand and resolve the above problems so as to not only to boost the food production in a convincing manner but also to ensure a sustainable agriculture production capable of supplying the future needs while conserving the natural resources and protecting the environment.

The present study is a modest attempt in the above direction which aims at emphasizing the need of devising agrometeorological management strategies through optimum use of information on climatic assessment and weather conditions so that the natural calamities hampering the food production are effectively curbed and managed. The subject is becoming increasingly popular on global basis and is receiving the attention of the agricultural scientists world over. Consequently, the need has also been felt for forging a joint strategy to devise means to boost agriculture production without trifling with the environment. Recently a joint workshop of Ohio State University, U.S.A., and the Punjab Agricultural University was held at Ludhiana (The Hindustan Times, Dec. 16, 1999) to discuss the possibility of a joint venture by the universities to chalk out strategies in agriculture research and development in the new millennium. In his inaugural address to the workshop Dr. R.S. Paroda, Secretary, Department of Agricultural Research and Education, Ministry of Agriculture, emphasized the need of changing the old crop pattern. Citing the case of Punjab, he said that the state could not continue to live on wheat-rice rotation and it has to diversify to other crops. Interestingly, the reason he assigned for this was not only to raise the yield but "to sustain natural resources and preserve the environment". A lot of research work is thus coming up to highlight the need of an agrometeorological approach to agriculture. But there is sample scope still left in this field. Hence, the need of the present study.

As for the plan and structure of the study, it has been divided into fifteen chapters, each dealing with the different aspects of the subject which is inclusive of the introductory and concluding chapters. In the introductory chapter the need of maintaining the pace of agricultural production to match with the growth of population has been emphasized. It also underlines the need of utilizing agrometeorological informations in the agricultural operations to the maximum extent. The second chapter seeks to stress the importance of regional agrometeorological resources. It also brings out the necessity of the optimum use of the regional agro-climatic resources and illustrates the same by way of crop growth simulating models. In the third chapter which is entitled as "Predictive Models for the Outbreak of Pests and Diseases", it has been comprehensively demonstrated that how weather forecasts can help the farmers in applying pesticides and fungicides on their crops as it can be known beforehand that which type of weather is going to cause what type of problems. In the fourth chapter the global agrometeorological scenario has been discussed wherein the problem of environmental pollution and its serious consequences have been discussed. In this chapter the problem of global warming which is the result of extraordinary high level of CO_2 concentration in the atmosphere consequent upon the large-scale deforestation and over-industrialization in the wake of growing population has been discussed. In the fifth chapter the need of forging alternative strategies to deal with inconvenient weather conditions has been discussed so that the crops are safe from them. The sixth and the seventh chapters deal with the various forms of agrometeorological forecasts and agrometeorological advisory services respectively to help the farmers to plan their various agricultural activities accordingly. In the eighth chapter it has been shown that how computer technology can help in managing and fighting meteorological disasters like storms, cyclones, earthquake etc. in the ninth chapter a more detailed treatment has been given to the problem of increasing level of carbon dioxide and its ill-effects upon the growth of the plant. In the last and concluding chapter the gist of the study has been given which really presents a very gloomy picture of the global environmental condition and highlights the need of conserving natural resources as well as saving the environment from getting polluted. It also flashes a note of caution to the people at large to avoid indiscriminate exploitation of the natural assets. The main thrust of the study, however, is that unless we are guided by the agrometeorological leads, it will be difficult for us to grow enough food to feed the population in the coming millennium.

The study is based on the data collected from numerous sources which have been invariably acknowledged, the details of which have been given in the respective references. I am grateful to all such authorities whose works and findings have been a beacon light for me during the course of the study.

Dr. R.S. Mishra
Dr. Raj Bahadur

Contents

1

Introduction

Agricultural operation is a very complex exercise which to a great extent is subject to the dictates of nature. The location, site, situation, and climate of a place are the major factors to decide as to which crop should be grown in that region and what appliances, methods and equipments can be used at the various stages of the agricultural activities. For instance, you cannot grow jute in Punjab or cotton in West Bengal, tea in Kashmir or saffron in Assam, apple in Kerala or coconut in Himachal Pradesh, so on and so forth. It is, therefore, rightly observed that "no region or country can develop on lines contrary to what its natural conditions permit"[1] (Dasharatha Sharma, 1966).[1]

This is, however, outrageous to believe that man has absolutely no role to play except toeing the particular line of action pre-ordained by nature. Had it been so, the large tract of desert in Rajasthan would have been left lying barren and not turned into a green paradise smiling with all sorts of crops and vegetables, flowers and fruits being grown on a large scale. Obviously, this all has been possible by human efforts and not by any compassion and favour bestowed by nature that the course of water was changed upwards bringing the same into the large arid zone of the Thar Desert in the form of canals. Consequently not only the life and economy of the region but also its climate and ecology have undergone a drastic transformation.

As a matter of fact right from the dawn of human civilization it has been a continuous and persistent endeavour of man to conquer nature. The greater is his control upon the nature, the more advanced and civilized he ought to be considered. In the beginning man probably tried to overcome the nature through trial and error method by his limited means, sometimes succeeding and sometimes giving in but the fight continued unabated. The same fight is still on but the scale and extent, norms and parameters, means and methods, aims and objectives have altogether changed to suit the changing pace and priorities of the time. This fight, when it comes to making strategies and schemes, manoeuvres and manipulations, in modern and scientific terminology, is known as research.

In this way, research on a large scale is being conducted in various spheres. But those conducted in the field of agriculture are far more vital and significant as they are directly linked with the most dominating basic need of man *viz.* the food, without which his very existence and survival becomes impossible. The rise in the world population mainly by bringing down of mortality rate on account of effective check on diseases including those having propensity and magnitude of epidemics, as also the fillip and booster the human health has had because of the availability of preventive and remedial medicines, have made it all the more a dire necessity to grow more and more food.

In view of the above, the agriculture sector has been receiving top priority at the government level ever since the country got its independence. Agriculture universities and research and training institutes have been established to impart specialized education and training in agriculture. Even at the school level agriculture is being taught as a subject. Researches have been conducted and are still underway in a wholesome manner to develop high yielding varieties of food grains. New methods and techniques have come to be introduced into various cropping operations right from the preparation of soil to the harvesting which includes the use of new and more effective variety of tools for tilling, hoeing, harrowing, weeding and winnowing etc. Besides, the use of quality manures and chemicals to increase and preserve the fertility of the soil, pesticides, fungicides and other medicines to ward off pests and diseases, seeds to ensure high yield, equipment for irrigation to avoid wastage of water, have come to be initiated, which has in fact gone a long way to revolutionize agriculture in the country bringing about what is popularly known as the Green Revolution.

But unfortunately in spite of this landmark achievement the Indian agriculture lags far behind many developed countries of the world. We have enough potential, being misused or still lying untapped, to help us better our performance in the field. Since our man power is not being properly utilized and the natural resources are being squandered away senselessly, the picture is likely to become far more bleak and obliterated unless some effective measures are adopted to correct the situation. The frequent occurrences of natural calamities like flood, drought, cyclone etc. coupled with environmental pollution, global warming caused by greenhouse effect and the likes, have deteriorated the situation to an alarming level. In view of this the talk of increasing the agriculture production to keep pace with the other developing and developed countries, even to sustain the level already achieved in that sphere has become a difficult task. Therefore, the need is to base agriculture on an eco-friendly foundation which will entail greater awareness of the weather as a pre-requisite. This will, on one hand, enable us to develop such varieties of crops and cropping pattern which suit our climate conditions most, while on the other it is likely to sere as a guide to various agricultural activities ensuring safety against the aberrant weather and its ill effects like flood, drought, storm, cyclone, earthquake, as well as the menace of pests and diseases. Thus an agrometeorological approach to agriculture is essential to boost the food production in a sustained manner.

Sustainable agriculture production is, in fact, the concept of renewal which envisages continued economic growth without destruction of natural ecosystems. Agricultural development programmes in India and elsewhere including cultivation of high yielding varieties over large areas, development of irrigation facilities particularly proper exploitation of groundwater resources, adequate and balanced use of fertilizers, adoption of plant protection measures and a well organized and systematic supply inputs with monetary assistance are all based on agrometeorology. Because air, water, land and energy which have tremendous effect on agricultural production system can be managed in accordance with agrometeorological information. In general, sustainable environment is linked with soil aeration, moisture availability, thermal and wind speed and direction, which cause increase in chemical toxicity, impeding crop growth and excessive transpiration loss at high temperature and wind speed, injury at the cellular level under cold conditions and uprooting in a windy condition, which can be minimized thorough proper agrometeorological management strategy.

In this way the application of agrometeorological techniques such as weather forecasting and modelling (predictive model for outbreak and spread of diseases and pests, predictive model for yield) can play a major role not only for increasing agricultural production but also for creating eco-sound environment, economizing inputs, minimizing the losses and wastage of energy, and maximizing the use of renewable natural resources. Therefore, an agrometeorological management strategy has to be made for the optimum use of information available on climate assessment and weather prediction, which is also environmentally sound, economically viable and capable of enhancing production.

2

Crop Growth Simulation Models and Regional Agrometeorological Resources

The primary objective of 'crop growth models' and the validation is to predict the growth and productivity of various crops, which is dependent upon several climatic, edaphic, hydrological, physiological and management factors (Jones and O Toole, 1987). The major factors affecting crop growth and development are radiation, temperature, water and nutrition availability (Nix 1987). In addition to this, other factors also determine productivity of crops such as varieties and their physiology, crop management operations vis-à-vis weather and soil. These operations vary from region to region in accordance with the peculiarities of the place. There are large areas where input use needs to be intensified whereas there are areas, particularly in the north-western parts of the country, where input use is already very high but unbalanced and untimely. Therefore, it is essential to evolve a more rational management scheme there including efficient use of inputs, particularly irrigation and nutrients. Crop management in such a production system is generally more complex, decision making is rather difficult, and economic returns are relatively smaller. In order to obtain better returns in such systems considerable managerial and informative skill on the part of the farmers, extension workers and researchers is required.

Hence a large number of experiments have been done earlier to study the effect of irrigation and or water on productivity of wheat (Aggarwal and Kalra, 1994). Multiple regression models have also been developed to determine irrigation and nitrogen requirements of crops (Kroentazer *et al.*, 1988). Besides, crop growth simulation techniques are recently developed to test the worth of numerous combinations *e.g.*

climatic variability x cultivar x regions x water availability x sowing time x other of the climatic variables, however, have a fixed value which vary from time to time. These computer based models have the additional advantage of taking less time in analyzing the effect of a dynamic climatic variability on crop productivity. Therefore, such models as WTGROWS and MACROS are being used increasingly for finding out an improved agriculture system which is both eco-friendly and economical (Nix 1987, Agrawal, 1992).

Table 2.1: Grain Yields (t/ha) of Wheat in Current Weather and Per cent Change in Response to Climate Change (425 ppm CO_2, 2°C increased mean temperature) in different Regions of India

Region No.	Potential Yield		Irrigated Yield		Rainfed Yield	
	Current Per cent	Change	Current Per cent	Change	Current Per cent	Change
>27	6.66	−3.85	4.89	+8.7	2.95	+28.6
25-27	5.84	−1.54	4.78	−4.4	3.34	−7.2
23-25	5.86	−5.6	4.18	−10.7	1.17	−19.6
20-23	4.18	−18.4	2.29	−18.3	0.51	−11.8
<20	3.69	−17.3	2.43	−21.4	0.97	−23.9

Source: Agrawal and Kalra, 1996.

Model output in Table 2.1 shows that the mean yields for current and changed climatic scenario (2°C rise in temperature and CO_2 level 425 ppm) according to latitude at irrigated and rainfed conditions in comparison to potential increase or decrease on the location specific to wheat growing areas. Therefore, it needs to adopt better managerial practices to stop where yields are going down (Aggrawal and Kalra 1994a).

Table 2.2: Simulated Grain Yield (t/ha) (GY) and Ground Biomass (t/ha) TDM grown Under Adequately Irrigated Condition for Observed and Forecast Weather of New Delhi and Ludhiana

	GY		TDM	
	A	B	A	B
Delhi	5.61	5.71	13.08	13.37
	6.61	6.64	15.51	15.66

A: Observed weather; B: Forecast weather.

Agroclimatic Zones

There is need to develop regional technology which can be adopted by real ground level farmers to give sustainable production. In this context scientists have developed some regional cultivars which produce maximum output in a given climate and soil which may be termed as agroclimatic conditional variety. A country like

India, which has diverse climate and soil type, needs efficient planning of cropping strategy. Hence, it is necessary to divide it into well-defined agroclimatic zones. The planning commission (Khanna, 1989) has divided India into 15 agroclimatic zones. Recently, the National Bureau of Soil Survey and land use planning of ICAR has, however, divided it into 21 zones on the basis of agroclimatic resources and soil condition (Sehgal *et al.*, 1990). These climatic variations are very important for planning and ensuring sustainable agricultural production. The fifteen zones which India has been placed in by the Planning Commission are as follows:

1. Western Himalayan Zone
2. Eastern Himalayan Zone
3. Lower Gangetic Zone
4. Middle Gangetic Zone
5. Upper Gangetic Zone
6. Trans-Gangetic Zone
7. Eastern Plateau and Hills
8. Western Plateau and Hills
9. Central Plateau and Hills
10. Southern Plateau and Hills
11. Eastern Coastal and Hills
12. Western Coastal and Hills
13. Gujarat Dried Zone
14. Western Arid Zone
15. Island Zone

Dr. M. S. Randhawa has, however, divided India into the following 5 agricultural regions on the basis of climate, crops and stock animals (M.S. Randhawa, 1980, 23-24).

1. The Temperate Himalayan Region: It has been further divided into two sub-divisions *viz.*, the eastern Himalayan Region, which includes Mishmi Hills in Upper Assam, Sikkim, Bhutan and Nepal; and the Western Himalayan Region which includes Kumaon, Garhwal, Himachal Pradesh and Jammu and Kashmir.

2. The Dry Northern Wheat Region: It comprises Punjab, Haryana, western Uttar Pradesh, western Madhya Pradesh, and parts of Rajasthan.

3. The Eastern Rice Region: It comprises Assam, West Bengal, Bihar, Orissa, eastern Madhya Pradesh, eastern Uttar Pradesh, and parts of Andhra Pradesh.

4. The Malabar Coconut Region: It comprises Kerala, the Western Coast strip, Karnataka, and the adjoining areas.

5. The Southern Millet Region: It comprises the Jhansi Division in southern Uttar Pradesh, central Madhya Pradesh, western Andhra Pradesh, western Tamil Nadu, eastern Maharashtra, and parts of Karnataka.

Naturally the agricultural activities in each zone because of its physical and climatic peculiarities have to be widely different, which calls for adopting crops and crop pattern accordingly to obtain maximum returns. Emphasizing upon the need of conserving natural resources while striving for enhancing the production, Dr. M.S. Swaminathan, Chairman of the M. S. Swaminathan Research Foundation (MSSRF), Chennai, says that in a populous country like India, where the per capita availability of land and water resources is going down, the need of the hour is a vertical growth in food grain production without endangering the ecological assets. To meet the food requirements of the growing population, there should be another green revolution with a difference, the aim of which should be to find out ways to improve food production, raise farmers' income and generate employment opportunities in rural areas through on-farm and off-farm activities, he opines. An integrated intensive farming system (IIFS), for example, with its components such as irrigated rice along with fish, other crops in mixed or rotational, for *e.g.* vegetable-fruit trees, and poultry, livestock in appropriate combinations in different agroclimatic zones, is a viable option that can provide both food security and livelihood to the rural and regional work force on an ecologically sustainable basis. (Venkatramani, 1955).

3

Predictive Models for the Outbreak of Pests and Diseases

Weather forecasts and meteorological advice and information can be very helpful in protecting the crop from the menace of pests and diseases as it enables to make appropriate strategies well in advance to deal with them. The life cycle, severity and threshold population of insects and diseases depend upon the prevailing weather conditions. Therefore, weather information and counselling, particularly regarding the effect of weather on the population density of pests and insects and also the outbreak of diseases on the crop, are very useful in predicting the exact nature of pests and disease likely to spread. Biopathogens are scanty in those regions where the climatic conditions changes along with the crop requirement, because pathogen and any necessary vector cannot grow there due to lack of favourable weather conditions. Thus, the development of insects and diseases as per the classical forecast model is dependent on patho- system *i.e.* a virulent pathogen is a susceptible host and subject to favourable environment.

In recent years, our effort to improve the ability to understand climate and nature of pest attacks to predict the development of epidemic has been more or less successful. The plant pathologists and meteorologists through coordinated efforts have been able to develop models for potential epidemics of some common and serious diseases such as BLIGHTCAST (Potato late blight), EPIDEM (Tomato and potato early blight), MYCOS (Mycosphoerella blight of Chrysanthemums), EPICORN (Southern corn leaf blight), EPIVEN (Apple Scab) (Agrios, 1988). Plant disease forecasting models give warnings to the farmers that patho-system is conducive for epidemics, so preventive spray application is need (De Weille, 1965). As a matter of illustration some models

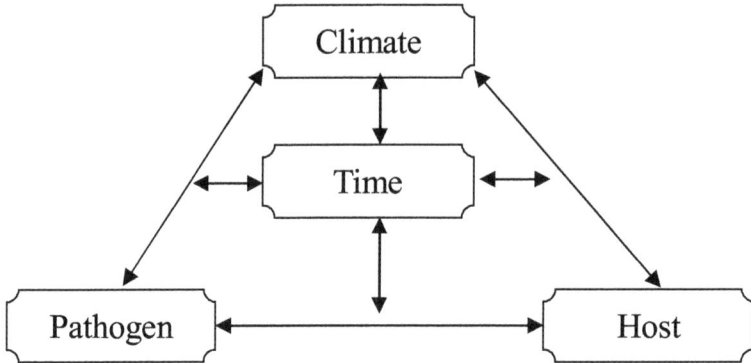

Climate

Time

Pathogen

Host

Figure 3.1

are given below:

1. Potato Late Blight Models

Reliable warnings based on relative humidity and minimum temperature observation at the 40 cm level, as also the normal synoptic weather observations from regular weather stations, can be flashed to forecast the outbreak of late blight in potato and tomato. The basis of such predictions is:

The average relative humidity should be 82-90 per cent for two consecutive days.

The minimum temperature should be 10°C at least for one day.

The maximum temperature should be below 19°C.

A cloudy weather should be more than two consecutive days.

Precipitation should be 1-2 mm a day.

2. Forecast of Brown Leaf Spot of Rice (Padmanabhan, 1963)

In the same way according to the Padmnabhan model heavy rainfall in September accompanied by a favourable temperature variation of 25-30°C followed by continuous cloudy weather during October and November, higher minimum temperature than the usual in November, low solar radiation in this month, are favourable to stimulate the formation of maximum number of conidia to cause epiphytotic.

3. Strips Rust in Wheat-Regional Models

Models developed by Coakley *et al*. (1984) to predict stripe rust on winter wheat are based on accumulation of negative degree days (NDD) in the winter and positive degree days (PDD) in the spring. The final disease severity can be calculated by NDD and PDD for the negative and positive departures of the daily average temperature from 7°C base. When there is less than 500 NDD, severe rust is likely to develop.

Table 3.1: Positive Degree Days and Prediction of Disease Intensity Index
from the Mean DI

Location	Mean of PDD Accumulated 1 April to 30 June	Mean DI	Difference DI from Pullman	Predicted Difference of DI from Pullman	NDDZ Residual
Pullman	429	4.5	00	0.2	00
Lind	607	3.8	−0.7	−0.6	−1.5
Mt. Vernon	444	5.0	0.5	0.1	1.1
Walla Walla	784	3.6	−0.9	−1.5	−1.2
Pendleton	713	2.7	−1.8	−1.1	−1.6

4. Northern Leaf Blight of Maize Forecast System

Disease develops easily at the temperature from 18-27°C. It is a conidial germination and infection occurs within 6 to 18 hours at this temperature and 90 per cent and relative humidity. Seven hours of humidity near 100 per cent and temperature above 15°C are found to be required for sporulation. The most of the conidia are released in the morning in the daylight hours which have the required moisture for germination. On the basis of such blight favourable hours (BFH) one can forecast epiphytotic disease.

Thus the nature of forecast required knowledge as well as the minimum skill of the forecaster, which depend upon the agricultural system as shown in the different crops for disease prediction. Detailed study of every cropping system is required for the identification of the nature of the patho-system which has a large impact on productivity. To assess the impact of adverse weather conditions such as dry/wet spells all the other factors which influence the outbreak of diseases have to be taken into account. Obviously, it entails an inter-disciplinary approach and a concerted effort on the part of the specialists in various fields comprising not only the meteorologists, plant pathologists and agricultural scientists but also the practising farmers, so that the management and cropping strategies of the farming system can be chosen for the maximum sustained productivity level.

4

Expert Support System for Global Climatic Change and Sustainable Agriculture

The global climatic scenario presents a very gloomy picture with the growing level of carbon dioxide in the atmosphere. Earlier, forests were considered as the permanent 'carbon sinks' but a recent report of the inter-governmental panel has demolished this theory (The Hindustan Times, Dec. 19, 1999). It has expressed apprehensions that due to the occurrence of a new phenomenon known as 'carbon dioxide fertilization' early in the next century forests planted to protect the planet from global warming may actually be contributing to it. In fact, this phenomenon is likely to take place because of the over-industrialization and the increasing density of population which has hopelessly destabilized the ecosystem and the process of respiration. If true, this is going to have serious repercussions also on the food front.

There is, however, an intense debate on whether the terrestrial biosphere (vegetation and soil combined) is responding to post-industrial human errors by emitting more carbon dioxide into the atmosphere or by sequestering it as organic carbon (C). There is no doubt that land clearing causes the oxidation of organic C to CO_2. Land clearing mostly by tropical deforestation is estimated to have been releasing CO_2 in the atmosphere to the extent of 1.5 to 3.0 Gt C Yr^{-1} in the late 1980X (Hougton, 1991). However, subsequent revaluation has reduced this to 1.7-1.0 Gt C Yr^{-1}.

The first estimate of the net global C release from the terrestrial biosphere, using a method based on measuring the rate of change in the atmospheric O_2/N_2 ratio (Keeling and Shertz, 1992), failed to discern whether the terrestrial biosphere overall was a significant source or a significant sink for CO_2. The range of estimate taken together with the deforestation estimate implies that the land, other than where deforestation is occurring, has a sink of C to the extent of 2.5-2.7 Gt C Yr[-1]. This estimate is similar to the so-called "missing C sink" in the global C budget (Watson *et al.*, 1990). According to the following table, identified source and sink relationship between emission of greenhouse gases *viz.*, CO_2, methane, chlorofluorocarbons and nitrous oxide, which is also increasing global warming have been estimated about 0.3 °C/decade over the next century.

A	Global Sources	Gt C Yr[-1]	Reference
i)	Fossil fuel burning	5.8-6.2	Houghton (1991)
ii)	Tropical deforestation	1.5-3.0	Houghton (1991)
	Total global source	**7.3-9.2**	
B	Global sinks		Reference
i)	Atmospheric increase	3.6-4.0	Watson *et al.* (1990)
ii)	Ocean uptake	1.3-2.9	Watson *et al.* (1990)
	Total sinks	**4.9-6.9**	

Missing C sink in global climate =   total global source – total global sink

7.3 to 9.2   –   4.6 to 6.9 = 0.4 to 4.3

Range of global missing C Sink is 0.4 to 4.3 CtC Yr[-1] local, national and world food supplies are very severe. Therefore, it is important to assess the effects of environmental conditions, associated with global climate change *i.e.* elevated CO_2 and temperature on the productivity of agricultural crops. The information derived from assessment of crop responses under experimental conditions may be included in crop simulation models and used to estimate the consequent effect of different combinations of factors and interaction with other conditions *e.g.* soil type and nutrition on the productivity of plants.

Table 4.1: Effect of Elevated CO_2 on Plant Growth Parameters in Mungbean and Maize Cultivars (pre-flowering stage)

Species/Variety	CO_2 Conc. (ppm)	Shoot Length (cm)	Leaf Area (cm²)	Leaf Dry Weight (g)	Shoot Dry Weight (g)
1) *Vigna radiata*					
(a)	350	18.33	59.73	3.67	3.25
	600	22.83	83.50	4.59	3.54
(b)	350	19.66	46.66	2.72	3.57
	600	25.20	64.42	4.03	3.55
CD at 5 per cent	Varieties	NS	8.93	NS	NS

Contd...

Table 4.1–Contd...

Species/Variety	CO$_2$ Conc. (ppm)	Shoot Length (cm)	Leaf Area (cm^2)	Leaf Dry Weight (g)	Shoot Dry Weight (g)
	Treat.	0.38	8.93	NS	NS
	Inter.	NS	NS	NS	NS
2) *Zea mays*					
(a)	350	51.21	74.23	3.48	8.95
	600	51.80	77.60	3.46	8.78
(b)	350	39.80	56.91	3.61	7.60
	600	40.50	57.16	3.66	8.08
CD at 5 per cent	Varieties	NS	11.23	NS	NS
	Treat.	NS	NS	NS	NS
	Inter.	NS	NS	NS	NS

Source: Uprety *et al.* (1996) Proc. Indian. Nat. Sci. Acad. (B62).

Table 4.2: Interactive Effect of Elevated CO$_2$ and Moisture Stress on the Harvest Index (per cent) in Brassica Species

Treat./Species	CO$_2$ Conc. (ppm)	Control (Irrigated)	Moisture Stress	Per cent Inc./Dec.
B. campestris	350	68.43	58.51	−14.50
	600	74.49	68.00	−8.72
B. juncea	350	71.59	34.92	−51.52
	600	72.00	50.94	−29.30
B. carinata	350	29.24	32.80	−11.68
	600	34.38	33.91	−11.68
B. nigra	350	25.06	22.73	−9.2
	600	26.21	24.59	−6.5

Source: Mishra, R.S. (1996) Ph.D. Thesis.

Effect of the rapidly increasing CO$_2$ concentration (currently increasing at 1.5 UL L-1 annum-1) have been examined for some of the most important food crops *e.g.* soyabean, mustard, mung and wheat (Uprety *et al.*, 1997, 1996), hut a few studies have shown good response in crops with a very long growing season, such as winter wheat (Kimball 1983).

The majority of these studies on plants have been conducted in controlled environment usually at reduced rotations and warmer temperatures as compared to the field conditions (Lawlor and Mitchell 1991). It has been suggested that a decrease in photosynthetic capacity in response to CO$_2$ enrichment is more likely to occur in controlled environments than in the field (Arp 1991). In any case, the response of plants to elevated CO_2 is very dependent upon other environmental factors, particularly water supply temperature and nutrition (Lawlor and Mitchell 1991). This is illustrated by the variability of crop responses to doubling of CO_2 concentration (Cure and Acock 1986) *e.g.* increases in wheat and mustard yield ranging from 5 to 37 per cent have been reported even under optimum water and nutrient conditions (Fisher and Aquilar 1976, Mishra 1996).

Like CO_2 concentration temperature has also pround effects on crop production. It affects the rate of organ development, respiration and senescence and altering the source-sink relations of plants (Farren and Williams 1991). Photorespiration increases in accordance with temperature, so positive interaction between increased temperature and elevated CO_2 concentration on photosynthesis is expected better in C_3 plants than C_4 plants (Long 1991, Uprety *et al.*, 1996). Increase in temperature upto $4^\circ C$ at the end of the next century has been predicted by global circulation models to corresponding increases of CO_2 from the current 360 ppm to 700 ppm. These conditions to present extremes but provide a baseline for testing the response of plants from which responses to less extreme conditions can be judged.

The complexity of the interaction among CO_2, temperature, and other environmental variables need to be examined for the quantification of plant responses to novel climate simulation models. However, these models contain approximations and empirical relationship for the regulation of there conditions as close as possible to those currently experience, and also expected in the near future in the field; *e.g.* natural day length and radiations. But it is difficult to do it in the field, therefore open top CO_2 chamber (Mishra 1996) and increasing CO_2 around crops by free air CO_2 enrichment (FACE) technology (Hondrey *et al.*, 1988) are needed. But it is extremely expensive in view of equipment and use of CO_2. In addition to this, temperature cannot be modified in this technology, so CO_2 x temperature interaction is impossible to study, except by exploiting temperature variability over the years. The FACE technology is useful in the large area of crop exposed to elevated CO_2. But open top CO_2 chambers are a practical solution to the problem of exposing crops in the CO_2 to altered atmosphere (Mishra 1996). Other than this, CO_2 also needs to modify the environment substantially increasing temperature and decreasing rainfall with the selection of suitable crop species for the sustainable agriculture production in future.

National and International Study on Climate Change

It is now widely accepted among the scientists that the effects of climate change on our natural ecosystem can be anticipated. But because great uncertainties are associated with the magnitude direction of the climatic change, it is not easy to foresee its effects accurately. The complexity and global nature of the problems connected with the biological effects of the climate change needs coordination among the various scientific disciplines as well as crops natural borders. Although biology of the sun is lagging behind the other natural sciences as far as natural resources are concerned, a significant increase is there in research efforts directed towards an improved understanding and predictability with respect to biological effects of climate change on crop growing environment. The growing number of research programmes in this field, however, makes it very difficult to consume results of individual scientist for the development of predictive model to understand future changes in climate and crop phenology. Hence scientists need to keep informations about current position of climatic change at the national and international levels for getting the primary objective of presenting the scenario of the global climate change and its impact on crop productivity. Therefore, there is need of great attention on the works performed by the Inter-governmental Panel on Climate Change (IPCC). In 1990, the IPCC produced scenario for future emissions of greenhouse gases, and these were interpreted in terms of future changes in greenhouse gas concentrations and global mean temperature on the crop productivity.

Use of Decision Support Systems

A formal system of accountability for climate change and its management decisions is basic to improve institutional performance for the crop production. Resilience and accountability constitute the basis for the responsive and adoptive management and need to be built up for planning and management of the effects of climate change on stage-wise crop growth system for sustainable production. A pre-requisite for such management is a formal, analytical, quantitative and interactive expert decision support framework which is needed to communicate at farmer level its use in agricultural practices.

Expert Decision Support System (EDSS)

Expert Decision Support System (EDSS) is an inter-disciplinary and interactive computer based system, which can help decision makers to use data. Data based computer models or heuristic models are used by EDSS to solve unstructured or under specified problems. In changed climatic scenario planning suited for the adoption of EDSS technology is needed. The EDSS enables to check the factors responsible for unfavaourable climatic change which affects the biosphere adversely and tends to impede production. It would also permit both quantitative and subjective evaluation of the decision which is implemented for protection of such misadventures as deforestation, fossil fuel burning and unlimited urbanization. Simulations of models are the essential components of the EDSS. These help to understand functions of the global climate change (GCC) resources to the natural input and output. It is well-known that existing models do not completely represent all these GCC resource real data. Thus there is need to link all the constituents which deal with the real GCC data in the decision process which is collected from field experience and in an interactive manner. These data are used as institution for subjective value judgements and balanced with the quantitative prediction of the models value before they are made available to the farmers for implementation.

Hence, it is a continuing need to improve the efficiencies of FCC policy for increasing agriculture productivity and staying safe from environmental crises. The increasing use of computers and communication systems, in the software and graphic technology of expert systems, has facilitated to assess the climatic changes fast and cheap. The computer expert knowledge and experience of policy making, management and operation in EDSS technology are key factors for its success, which provide a tool for building accountability into global climate change and its management system, because EDSS is based on a formal and critical framework that can allow retracing the pathways for decision making.

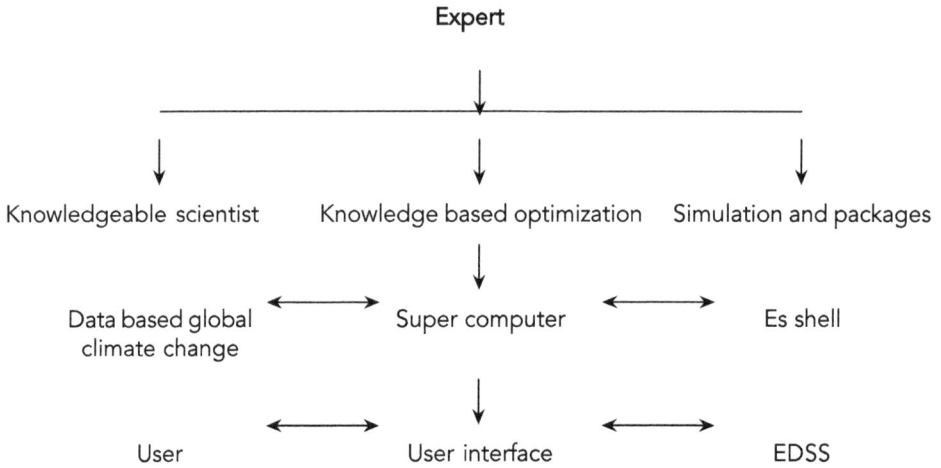

Expert

Knowledgeable scientist   Knowledge based optimization   Simulation and packages

Data based global   ⟷   Super computer   ⟷   Es shell
climate change

User   ⟷   User interface   ⟷   EDSS

5

Alternative Strategy against Aberrant Weather Conditions

India has diverse climatic conditions which have wide ramifications on agricultural production. As for instance, in the east we have the highest rainfall zone of the world while in the west there is the driest region in the form of the Thar Desert extending from north Rajasthan to the Runn of Kutch in Gujarat. Hence there is a need to adopt improvised technology to suit the weather conditions of a particular region for obtaining the sustainable level of production. We know that the aberrant weather in the form of unusual rainfall in the *Rabi* season badly affects the crop, growth, particularly when the rain occurs immediately after sowing. Likewise, *Kharif* crops are likely to suffer heavy damage due to rains during its maturity phases. This also results in delayed sowing of the Rabi crops like mustard, pea, gram etc. Besides, the adverse moisture stress during the early part of the crop growth and excessive rainfall during the flowering period hampers the seed setting of *Juar, Bajra* and *Til* resulting in poor yield. On the other hand, focusing further on the impact of adverse weather events it is suggested that the incidence of diseases and pests are more in the cases of groundnut, generally sown after the occurrence of the rain in July and harvested at the end of October. The adverse weather events cause major disease and pest attack in groundnut in Karnataka due to long dry spells which are also associated with incidence of leaf minor attacks, while the wet spells are associated with incidence of crown rot, late tikka disease and collor rot/(Godgil 1997). Effects of these adverse weather conditions on the crop productivity need to be analyzed.

Therefore, next we shall consider as the how climatological information of adverse weather events can be used by the farmers to decide about the remedial measures worth taking in the given situation. A practical prescription for remedy would undoubtedly need to cover the entire season and an elaborate analysis of crop situation. We, however, wish to illustrate here only the theoretical framework for making the appropriate decision.

Taking into account the rainfall data of 120 years (1870-1990) of the country, we notice that during this period drought occurred 425 times (72 times severely) (Hindu 1995). These are the aberrations of onset, continuity, intensity, volume and withdrawal pattern of rainfall that trigger drought like conditions. Long term analysis of rainfall has also revealed that the annual precipitations across the regions are decreasing. In other words, regions with low to medium rainfall covering an area of nearly 70mha are subjected to moisture deficiency and are also facing big storms. Besides that, high intensity rains produce volume of water beyond the intake capacity of soil. As a result a sizable portion of rainfall is lost via runoff from the area which is more than 30 per cent of the annual precipitation by runoff, and this exacerbates soil erosion. Unabated erosion knocks off nutrient rich topsoil, leaving behind sterile subsoil. More erosion occurs from bare undulating soils in general and impervious black clay soil in particular. If runoff is reclaimed, it can assist in building up permanent water resources for saving crops when they are facing stress due to long breaks of monsoon. Projections made by CRIDA scientists show that the existing potential of harvest 6.36 mha precipitation is equivalent to runoff at the farmers; field level in a region receiving 1000 mm rain per annum. The runoff, if harvested, can provide two life - saving irrigations of 5 cm each to more than 60 per cent of the total rainfed area (Katyal, 1995). Small and marginal farmers can conserve water through vegetative barriers like 'kush' though it has hardly shown any distinct superiority over other vegetative barriers generated from anjan grass; lemon grass; broom grass; kuntze and subabul. These are conventional bunding techniques. Vegetative barriers or even field boundary bunds, which are reinforced through simple land treatments, can maximize water soaking in the field profile. This is the basic rule to build-up *in-situ* moisture conservation, *via* stretching the "infiltration opportunity time".

Tillage also makes soil surface more permeable and fit for a high rate of infiltration. Deep tillage (25-30 cm) opens up the hard soil layers and accelerates penetration rates. If soil surface is loosened with pre-monsoon tillage, a strong foundation for rainwater conservation is built. The primary tillage also controls weeds and eliminates soil-borne diseases. However, different soils respond differently to tillage operations.

Interculture operations eliminate weeds and also give rise to ridges and furrows. Each furrow acts as a mini rainwater trap protected by two ridges through which up to 30 cmm ha additional water resource can be generated in the field. This strategy can ensure crop survival at least for two weeks should and unusually prolonged dry spell occur. Apart from runoff, evaporation also siphons a large amount of rain water (up to 60 per cent). Therefore, needs to use harmonizing agricultural practices for the conservation of rain water against loss through evaporation which is essential for sustaining growth when confronted with long dry spell (Hindu, 1995). Mulching, which is a man-made device, is done by covering soil surface with organic residues or by stirring the soil surface with inter-culturing implements leading to the arrest of evaporation. Seepage eradication is also necessary for soils with high percolation rates. Several alternatives such as cement brick lining plastering with dung and soil mixture have been examined for seepage proofing. But a short-term surface storage which encourages groundwater recharge with an apparent control of evaporative losses is more acceptable to the farmers. Date of commencement of monsoon, its stay time and withdrawal of rainy season are the basic need for the planning of agricultural operations and selection of crops and varieties which can withstand the aberrant weather conditions. In fact, unpredictable swings in rainfall behaviour do not allow uniform and universal recommendation on crop alteration for prevailing weather. Therefore, some farmers also do not mix cropping to avoid total crop failures. Further research is needed in this area for designing more profitable and stable intercropping options depended on temporal and spatial rainfall distributions. Crop combinations exhibiting minimal competition for light, water and nutrients have been recognized *e.g.*, pearl millet-clusterbean, groundnut-pigeon pea, sorghum-pigeon pea, upland rice-green gram intercropping suits where annual rainfall is 400-1200 mm (Katyal 1995).

Table 5.1: Rainfall Variability Across Meteorological Subdivisions 1990-1994

Rainfall	1990	1991	1992	1993	1994	Average
Excess/Normal	21	10	14	18	17	16
Deficit/Scanty	14	25	21	17	18	19
Total	35	35	35	35	35	35

I: No. of subdivision; Cv: 36 per cent; SE: 6.

Source: Economic News and Views 2 (3), Dec. 1994.

In the mid-crop period, if adverse weather events such as cloudy weather, frost, fog, rain, low temperature, high temperature etc. occurred, thereby increasing pest and disease occurrences, there is need for adopting remedial methods such as application of pesticides, fungicides and frost protection measures. In this case farmers cannot decide themselves whether or not to spray pesticide, fungicide on the basis of climatological information on the probability of pest and disease outbreak. Therefore, there is need to take the services of an experienced and skilled forecaster to advise on what remedial methods be adopted to avoid impending losses in the yield (Mishra and Padmakar 1997 unpublished). In addition to agro-forestry, agri-horticulture and silvipasture are typical examples where alternate strategies can minimize losses due to adverse weather and increase the productivity in a sustainable manner to meet the human needs.

6

Development of Agrometeorological Forecast

The weather forecast predicting rainfall, sunshine hours, temperature, wind speed etc. are imperative for the successful planning and management of agricultural operations so as to achieve optimum and sustained food production. In other words, the weather conditions need invariably be taken into account before commencing the activities like sowing, irrigation, pest/disease management, harvesting and storage. The weather forecast being used in the management and handling of the agricultural operations requires weather data of the immediate past and also that of past 30 years of the area (Kunkel *et al.*, 1990). The historical weather data must be obtained either from the control data system or from an independent data source operated for specific need. The weather data as indicated above must include maximum/minimum temperature, precipitation, total evaporation, wind speed/direction, humidity, cloud cover, sunshine hours etc. Before using the meteorological data for weather forecast, there is need to check quality of the data with the standard quality control method prescribed by world meteorological organization and Indian Meteorological Department, Pune. Because meteorological observers can commit mistakes in making the observations, writing or sending data to National Data Collection Centre (NDCC) for comparing and processing is essential. After completing all these necessary formalities, agricultural weather forecasts should be issued for being utilized in crop growing activities including day to day farming operations and future planning for the sustainable agriculture production.

Forms of Weather Forecast

Weather forecasts for agriculture can be divided into three categories viz,.

1. Short-range forecast valid upto 48 hours.
2. Medium-range forecast valid for 3 to 10 days.
3. Long-range forecast valid for more than 10 days or even month or season.

Short-Range Forecast

The short-range weather forecast is based on synoptic situation prevailing at the time of forecast and is valid upto 48 hours. In the synoptic approach, a human forecaster attempts to predict the future changes in the state of the atmosphere from the initial state using his theoretical knowledge and experience. The forecaster analyses current data available at various heights of the atmosphere and tries to match it with the past similar situation, assuming that the future behaviour of the present situation will be more or less analogous to the past situation. The correctness of the perditions made as such, however, depends upon the skill and experience of the forecaster. Hence the method is subjective but very useful in short-range predictions.

In addition to the synoptic method, numerical weather prediction method is also used for short-range forecast. The use of super computer technique is also being made for short-range weather forecasting to facilitate agricultural operations. Short-range forecast helps farmers in the implementation of instant agricultural works like sowing the seed or watering the crop in accordance with the dictates of the weather. To be more specific, at this time if the forecaster predicts rainfall in a day or two, the farmers stop the sowing and irrigation opeations in view of the impending rainfall. Similarly, if the farmer is about to spray pesticide or fungicide in the crop to ward off pests and diseases and in the mean time weather forecaster forecasts heavy rainfall and dust storm in the area, the farmer will postpone his spray programme till the availability of the favourable weather condition, otherwise it is likely to cause double loss, one monitory in the form of the cost of the chemical and second environment as the chemical cause pollution. The short-range forecast is thus very useful which is also called as operational forecast.

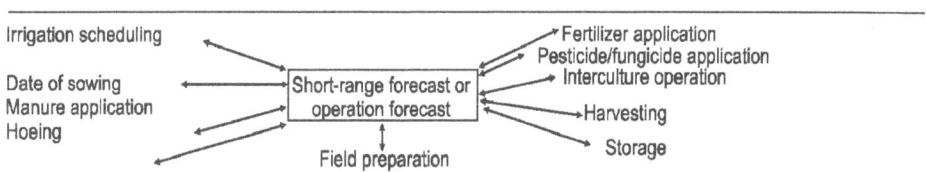

Irrigation scheduling		Fertilizer application
		Pesticide/fungicide application
Date of sowing	Short-range forecast or	Interculture operation
Manure application	operation forecast	Harvesting
Hoeing		Storage
	Field preparation	

Medium-Range Forecast

Medium-range weather forecasts for different agro-climatic zones valid for a period of 3-10 days are very important for agriculture. Medium-range forecasts are successful due to the advent of super computers and very efficient numerical schemes.

Numerical weather prediction system has become a promising technique of weather forecasting especially for the medium range forecast. In this technique hydrodynamic equations governing the atmospheric motions are solved by numerical methods using high speed computers. This computer based weather forecast method being objective has a good success rage outside the tropical belt. Therefore, medium-range weather forecast is important for agricultural operations and management planning in multifarious ways including water management, pest/disease management, seed and fertilizer management. Thus agrometeorologists thought their medium range weather forecasts can direct the farmers for various agricultural operations such as sowing, irrigating, spraying, fertilizing etc. at least 10 days in advance which enable them to make necessary arrangements for seeds, fertilizers, pesticide and fungicide to get maximum and sustainable agricultural production. This forecasts also gives beforehand an idea for crops to be grown and protective measures to be taken. Hence this forecast is called as planner forecast.

Medium-range forecast

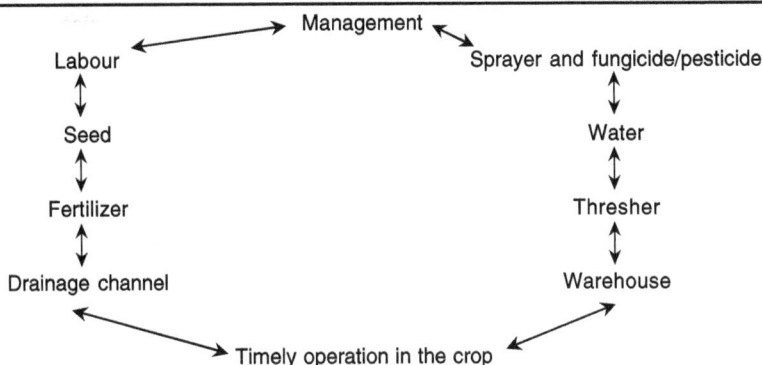

Long-Range Forecast

Long-range weather forecast is issued more than 10 days, a month, or sometimes even a season ahead. These outlooks are usually expressed in the form of expected deviation from the normal weather conditions. Long-range weather forecasts are issued on the basis of statistical approach. A weather forecaster tries to correlate one weather element with another by studying the past weather records. Useful relationships can, in fact, be established relating the occurrence of one weather parameter with another or a number of other weathers. In Addition to statistical method, other methods like numerical weather prediction and direct model output are also used for long range weather forecast. These forecasts are used at all levels—national, state, and individual. For instance, weather forecast is for a heavy rainfall, drought, cold/heat wave, hail/thunder/dust storm, or high/low humidity, the national/state policy makers make policy accordingly. That is to say, if the forecast is made for a heavy rainfall and the resultant flood which is likely to inflict heavy losses to crops, livestock's and human life, the policy makers at various levels get ample time to plan defensive strategies thereby minimizing the losses if not altogether averting them. Similarly, if the paucity of rain is predicted and there is likelihood of drought, government and farmers can adopt suitable water management and cropping pattern for the minimization of the losses. In the case of probable cold and heat wave conditions, farmers can be advised to use temperature resistant crops and varieties. Likewise, the thunder and dust storms can be fought by making windbreakers along with crops and also using deep-rooted crops which can sustain adverse effect of the storm. Weather forecast is also useful for fighting the menace of pest/disease in a particular crop or variety thereby saving it from the probable losses.

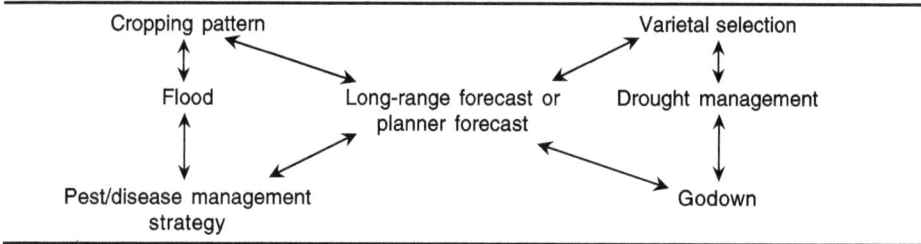

Cropping pattern		Varietal selection
Flood	Long-range forecast or planner forecast	Drought management
Pest/disease management strategy		Godown

7

Agrometeorological Advisory Services and Co-ordination Research

Agrometeorological advisory services (AAS) are the inter-disciplinary activities of Agronomy, Soil science, Plant pathology, Entomology, Horticulture and Meteorology. This multi-disciplinary scheme can enhance the yield on account of the eco-sound environment envisaged therein. The inter-disciplinary co-operation liaisons are well conceived because of the common goal of translating the theoretical knowledge of the experts into practice at the farmers level. As to how even a little transgression of this principle can cause irreparable losses can be understood by the example of the overdose of pesticides being applied all these years, which has caused serious environmental pollution. Keeping this in view, scientists of plant protection are busy promoting integrated pest management (IPM), which is environmentally sound, economically viable, and socially acceptable. For this purpose, the government reoriented its policy on plant protection during 1985 by adopting IPM as its cardinal principle and main plank of plant protection strategy in the overall crop protection programme. It has taken positive initiative in the promotion of IMM, such as human resource development though a three tier programme consisting of a seasonal/long-term training programme for subject matter specialist (SMS), establishment of Farmers Field School (FFS) to train farmers. In the FFS demonstrations for adoption of field tested IMP technology, policy to promote *neem* based pesticides and bio-pesticides, and phase out the use of hazardous pesticides, are conducted.

Table 7.1: The Salient Feature of the Case Study Before and After the
Establishment of Farmers Field School (FFS) for IPM

Sl.No.	Parameters	Before FFS	After FSS
1	Agro-ecosystem analysis	15.8 per cent	100 per cent
2	Decision-making in crop management	27.6 per cent	99.5 per cent
3	Reduction in pesticide use	–	78.0 per cent
4	Productivity (t/ha	4.7 per cent	5.3 per cent

Source: Technical note of V. Ragunathan, 1995.

Agrometeorological Advisory Bulletins (AAB)

The periodical agrometeorological advisory bulletins are not the handiwork of meteorologists but the team work of the experts belonging to various disciplines as agrometeorology is not a field of meteorology alone. They deal with the relation between crop and weather which no single research organization pertaining to a particular branch can really accomplish. The agrometeorological advisory bulletins which are aimed at ensuring sustainable agriculture production serve as an operational tool for farmers' day-to-day activities by issuing weekly, monthly, and seasonal forecasts for long and short-term planning.

The format of agrometeorological advisory bulletins includes synopsis of the past week's weather conditions and forecast of three days to one week or even that of a month or season, with the details of crop situation and operation. Meteorological forecasts put emphasis on crop operation in the real operational fields. For this purpose crop weather interactions are analyzed to give suitable advice for agricultural operations at least one week in advance so that the farmers can plan their activities accordingly and decide about sowing of crop variety, time of irrigation, and fertilizer, pest and disease control measures.

Language of agrometeorological advisory bulletins should be local, which could be easily utilized by the farmers in the day-to-day field operations. The dissemination of these bulletins are also accounted by time and ways. Therefore, these bulletins need to be broadcast by Radio and TV preferably in the morning and evening when the farmers are available in their homes. Personal contact between farmers and scientists will be very effective to help for solving such problems of the farmers on the spot as are not included in the bulletins. Thus mass communication media including print and audio-visual are very useful, being quick and relatively cheaper, and can serve as a very effective tool for communicating weather reports.

Co-ordinated Research and Development

Co-ordinated research is a continued and major need for improving the lot of Indian Agriculture. Even if the country has realized the goal of full self-dependence in the field of the production, without co-ordinated research and development programme, it cannot be stable for long due to the degeneration of natural resources such as soil, water, and environment. Hence it the prime need of the time to have a multi-disciplinary approach to the problem and concerted effort for research and development forging proper working cooperation between the research worker and the farmers for the effective implementation of various advanced programmes and technologies. This has been one of the one the major lacunas in the past extension efforts that the effective participation of farmers in the "technology transfer programmes could not be obtained. Farmer's participation is, in fact, important in research, planning and technology development and its transfer in the field. Water conservation technology like watershed development is need to community participation for its success. Participatory technology development (PTD) should also be used as an important tool to facilitate quick diffusion of technology to the need of community. Social science research should, therefore, receive high priority both for assessment and evolving proper solutions to the problems of production and conservation. Documentation of the indigenous technological knowledge (ITK) and the traditional wisdom of the farmers shall be the corner stone of such efforts. Thus farmers will be a focal point in the problem identification, research planning and technology development to suit the future needs.

Table 7.2: The Estimated Potential Availability of Nutrients from
Three Major Groups of Recyclable Resources for INM

	Total Potential	*For Land Use*
Crop residue	5.6-8.7	1.7-2.6
Animal dung	3.4-5.7	1.0-1.7
Human faeces	1.51.8	1.2-1.4
Total	10.5-16.2	3.9-5.7

Source: The Hindu Survey of Indian Agriculture 1995.

Effective diffusion of available technology also needs the involvement of non-governmental organization (NGOs). These NGOs should particularly be deployed in the areas like large programme implementation *viz.*, watershed construction, adoption of agro-advisory recommendations etc. where community action is required. Therefore, formal and informal linkages have to be established among scientific organization, NGOs and farmers to evaluate different models of technology transfer and the identification of most suitable model for the application in other areas. Besides, the farmers, research scientists from different disciplines *viz.*, agrometeorology, agronomy, soil science, entomology, and plant pathology, should also make co-ordinated efforts for research planning and implementation in the farmers' field. One of such co-ordianted programmes is the integrated nutrient management (INM), which is limited to the crop yield in dry land regions. Crop failures due to aberrant weather has always restricted the farmers from using recommended levels of fertilizers use of inorganic and organic nutrients are suggested in the INM package, but the unavailability of adequate amount of organic materials like crop residues, FYM and green leaf materials has been a major constraint. Raising, transporting and adding green leaf materials are still not cost effective. Therefore, strategies to raise organic matter on a non-competitive basis either through bund farming or raising a fast growing legume either before or after the normal crop season should be pursued.

This will also help in utilizing off-season rainfall. Such co-ordinated and detailed research on water and nutrient is needed also to regulate the fertilizer application. Research on the use of bio-fertilizers needs to be popularized for developing better strains, suitable for rainfed agriculture and identification of the factors which govern the spore inoculation responses in the crop.

From technical, economical, logical and environmental considerations, the best course is to practice integrated nutrient management (INM). This will bring about a harmonious and integrated use of minerals, fertilizers, organics, bio-fertilizers and recyclable wastes and is likely to stand the test of time. As for the INM package, it should be technically sound, economically attractive, practically feasible, socially acceptable, and biologically eco-friendly, so that this technology can be adopted in any field through a co-ordinated attempt.

8

Meteorological Disaster Management through Computer Technology

The weather forecasts are not limited only to recording temperature and predicting rainfall or giving prior idea of the outbreak of pests and diseases. They are now also being used to warn people of disasters like floods, droughts volcanic eruptions, forest fires and earthquakes. With the help of satellite images we can gather the information of the cyclones no sooner than they are formed in the ocean. Indian Space Research Organization (ISRO) is also going to launch weather observation satellites with more than 330 instruments. With the help of them it will be possible to monitor developing weather systems including forecasting drought and assessing the risk of landslides. Besides, Synthetic Aperture Radar (SAR) is helpful in pinpointing earthquake epicenters. Global Positioning Satellite (GPS) placed in high earthquake risk regions, in the same way, can detect ground movements that may be a precursor to an earthquake. There are also operational environmental satellite censors to detect and monitor wild fire which due to heavy smoke it is difficult to detect from aircraft (cf. The Hindustan Times, Dec. 19, 1999). It is thus obvious that meteorological device goes a long way in predicting and managing natural disasters like cyclones, earthquakes, wild fire etc.

Among Asian countries India ranks number one in the frequency of meteorological disasters. Kandla and Orissa cyclones are the latest in the long series of disasters striking India at regular intervals. Other than cyclone, dust storm, flood and drought are the other major disasters. Which cause the loss of crores of rupees annually? The computers have, in fact, come to help in planning and managing such disasters in a better way.

Meteorological disaster management planning has four stages *viz.*, Preparedness (including prevention), Response (rescue) Relief and Recovery (rehabilitation). At each stage the response mechanism and response time have to be planned. Much of the meteorological disaster planning is based on common sense and the experience of past disasters. Therefore, it is essential to have a detailed knowledge of the history of earlier disasters for the identification of an appropriate action plan and agencies that may undertake those actions *viz.* the resource inventory. This inventory will be successful only when the directory is kept up to date storing all the informations in the computer.

Another important step is to prepare a risk and vulnerability analysis for each area, village, district or state. The government of Maharashtra has used the latest techniques for charting out the disaster management planning to neutralize disaster involving a large area and a huge monetary loss. In the meteorological disaster management the role of geographical information system (GIS) is very significant. GIS is a tool derived from sophisticated computer technology, which allows synergizing of map data and tabular data/print data in the most efficient manner. Data base technology, which now encompasses both spatial (maps) and non-spatial (tabular and attribute) data with concurrent advancement in the methodology to store, organize and retrieve the data, is a factor that has facilitated the development of GIS. Commercial availability of data has also provided a new thrust to information processing and decision making. The rapid progress in the hardware technology in producing efficient computers ranging from the PC platforms to the most recent workstation has provided a boost to the desktop planning process.

Table 8.1: Risk Assessment and Vulnerability Analysis in Maharashtra

Disaster Type	Disasters of Maximum Risk	Vulnerability
Earthquakes	Latur, Beed, Purbhani, Nanded, Nagpur, Nasik, Thane, Satara, Pune, Sandi, Ratnagiri and Mumbai	High
Floods	Nagpur, Bhandra, Chandrapur, Gandchirali, Wardha, Mumbai, Ratnagiri, Pune, Sholapur and Nanded	High
Flash flood	Nagpur, Bhardha, Chandrapur, Gadchirali, Wardha, Parbhani, Ahmadnagar, Pune, Sholapur, Dhule and Nasik	High
Cyclones	Mumbai, Thane, Raigad, Ratnagiri and Sindhudurg	Low
Epidemics (Water borne disease)	All districts in Maharashtra	Medium

Other important source of meteorological information is the Remote Sensing (RS) observations from the satellite to provide data on the earth and its natural resources in a spatial format. The remote sensing data offer a synoptic view and large area coverage, which helps in obtaining the proverbial bird's eye view of the features. The integration of information derived from remote sensing data and surface data with other data sets, both spatial and non-spatial formats, provide tremendous

Table 8.2: Relative Intensity of Hazards Faced by Selected Countries in Asia and the Pacific

Country	Flood	Earthquake	Drought	Volcano	Cyclone	Landslide	ISU Name	Fire	En.	Civil Strife	Frost	Accident
Australia	S				S			S				
Bangladesh	S	S	S		S	L	L	L	M	M	M	L
China	S	S	S		M	L		M	L			L
Cook islands	L	S	S		M	L	M					L
Fiji	S	M	M		S	S	S			M		
Hong Kong	L	M			M	M		M	M	M		
India	S	M	S		M	L		M	M	M		L
Indonesia	M	S	M	M	L	L	L	M	L			
Laos	M		L					L				
Malaysia	M					L						
Myanmar	M	S	M		M		S					
Nepal	M	M	M			M		M	M			
Pakistan	S	S	M		L	L		L	L		L	
PNG	S	S	M	S	L	S	S	L	L	L	L	
Philippines	S	S	M	M	S	M	L	S	L	M		L
Solomonts	S	S	L	S	S	S	S	L	L			
Sri Lanka	M		M		M	S		L	L	S		
Thailand	S	L	S		M	L						
Tonga	M	S	M	S	S	L	S	L	L			
Vernatu	S	S	L	S	S	S	S	L	L			M
Vietnam	S	L	L		S	L						

potential for characterization and analysis of earth surface resources. The information derived as such needs to be understood carefully in relation to socio-economic situation for planning and development to obtain sustainable crop production and economic growth.

Basically GIS is a computer-based system designed to store edit, display and plot geographical data on computer. It has capability to make quick and unbiased decision which includes distance, direction, adjacency, relative location and other spatial concepts. The computerized data constitute a comprehensive digital data base, containing information about various resource fields such as land, water, vegetation and socio-economic details. This can be tapped as per the requirements to create information system such as Land Information System (LIS), Water Information System (WIS), and Disaster Management Information System (DMIS). During the meteorological hazard analysis phase, GIS can provide incidence maps and thematic maps. It can provide a map and tabular report for shelter inventory analysis, identifying the location, characteristics of shelter within the district/tehsil level. The detailed logistics information with greatly help to Government and farmers both for fighting the meteorological calamities as given below. It also facilitates the analysis of various hazards through map overlays such as potential flood zones etc.

At the response stage, it helps to adopt emergency vehicle routing. For example, the information on the location of severe dust-storm can be juxtaposed with the information on the availability of the neutralizing equipments, available closest to the scene of disaster. Their movement can also be used as an effective tool in recovery phase, using the GIS tools. The DMP prepared by the government of Maharashtra has arrived at the risk and vulnerability analysis for all districts.

Armed with an effective communication link up, through the latest technology and data base created using GIS, the initial knowledge of the state government would depend on the quality of the data that forms the inputs. It is also clear that the state government also proposes to establish the rural networks with the NGOs through training and simulation programmes. A very senior administrator said in the final analysis, "If you are a good administrator, you do not need a manual or plan; if you are not, the best plan will remain a document that adorns your book shelf. DMP involves a commitment in the form of detailed drills and exercises. How many of us have seen this happen", he asks.

9

Plant Responses to Elevated Carbon Dioxide Concentration

Now we shall see a little bit more comprehensively as to how plants respond to the elevated level of CO_2. The concentration of carbon dioxide (CO_2) in the atmosphere has in fact considerably increased which as pointed out earlier (Chap.IV) has assumed an alarming dimension. This increase shows a rise from pre-industrial value 270-280 ppm (Bilin *et al.*, 1986) to the current level of 360ppm (Cullato 1995). The rate of its enhancement is also accelerating from 1.5 to 1.8 ppm per year (Houghton *et al.*, 1990). This change is very rapid on geological time scale due to the increase in world human population and its increasing economic activities (Keeling 1986). This leads to the consumption of fossil fuel, destruction of forest and rise in rural industrialization (Mishra *et al.*, 1996) thereby releasing stored carbon by oxidative process and reducing storage of carbon in the standing vegetation. Consumption of fossil fuel will continue to rise during the foreseeable future (Keeling *et al.*, 1989) owing to which the current concentration value of the CO_2 might be doubled upto 600 ppm or even more in the 21st century. This exponential rise in CO_2 may create a challenging situation for the present-day crop varieties.

Most of the crop cultivars selected for maximum productivity have CO_2 resistance only upto 310-320 ppm and their responses to elevated CO_2 have been studied to 600ppm. As a result of the accumulation of CO_2 and other radio-active trace gases in the troposphere, a shift in the climate (temperature and rainfall patterns) has been predicted.

In these references if the climatic changes take place, plants will be directly affected by higher CO_2 concentration. Virtually all studies to date have shown that the high level of CO_2 enhances crop growth and minimizes the same types of stress (*i.e.* drought) (Uprety *et al.*, 1995, 1996; Mishra *et al.*, 1996) and substantially increases yield and yield attributes (Uprety *et al.*, 1996; Kimball *et al.*, 1990; Bazzaz 1990).

However, since individual species respond differently, comparative shifts might lead to alterations in the composition, structure and function of crop phenology. Much of our data has been derived from controlled growth and open top chamber (OTC) in the field. Now FACE air CO_2 enrichment systems are beginning to come on line (Mishra *et al.*, 1996; Uprety *et al.*, 1992). Here in the present study the effect of elevated CO_2 on the crops with emphasis on photosynthesis, water use efficiency, plant growth, dry matter, yield and yield components have been presented.

Photosynthesis

Elevated CO_2 enhances the rate of photosynthesis both in C_3 and C_4 plants. But this enhancement is greater in C_3 than in C_4 plants. This is caused by high CO_2 which reduces the competition from O_2 for Rubisco and increases its activation which reduces photorespiration (Bazzaz 1990; Goudriaan *et al.*, 1990; Uprety *et al.*, 1995). In the case of C_4 plants, photorespiration is much less affected by elevated CO_2 because photorespiration is suppressed by high CO_2 concentrating mechanism (Poorter 1992) as C_4 species are quickly saturated when CO_2 concentration rises, while in the C_3 species photosynthetic rate increases as the CO_2 concentration rises across the range of several hundred ppm (Allen *et al.*, 1990; Gifford and Morison 1985; Huber *et al.*, 1984; Radin *et al.*, 1987). Several early studies on the response of plants to elevated CO_2, examined and given in Table 9.1, show that it differs from species to species and plant to plant. Other investigators have also observed that elevated CO_2 increases photosynthesis when plants are first exposed to open top CO_2 chamber after that the rate of photosynthesis declines within the period of a week (Drake 1992). Decline in the photosynthetic rate is not well understood but the reasons for it have been proposed such as decline in carboxylation efficiency which may be caused by a decrease in the amount and activity of Rubisco (Fetcher *et al.*, 1988; Sage and Pearcy 1987; Sage *et al.*, 1989). Suppression of sucrose synthesis by an accumulation of starch (Ginn and Mauney 1980; Walker 1980) inhibition of the triose P-currier due to reduction in the activity of sucrose phosphate synthetase, the limitation of day time photo-synthate export from sources to sink (Delucia *et al.*, 1985) and insufficient sink in the plant are also some of the reasons for it (Koch *et al.*, 1986). Thus the effect of elevated CO_2 on photosynthetic capacity, still uncertain by the consideration of sink activity (*i.e.* carbon partitioning among the various plants organs), may help these findings (Cure *et al.*, 1987, 1991; Herold 1980). The recent studies of Arp (1991) and Thomas and Strain (1991) have shown a strong correlation between pot size and photosynthetic capacity, which suggest a need to carefully consider root growth and water availability around the plants for the examination of enhanced photosynthetic capacity due to elevated CO_2.

Table 9.1: Effect of Elevated CO_2 on Photosynthesis

Crop	CO_2 Concentration	Photosynthesis (μ mol CO_2 m^2s^{-1})	Reference
B. campestris	350	17.8	Uprety *et al.*, 1995
	600	21.8	Uprety *et al.*, 1995
B. nigra	350	16.2	Uprety *et al.*, 1995
	600	20.3	Uprety *et al.*, 1995
B. juncea	350	22.8	Mishra *et al.*, 1996
	600	24.5	Mishra *et al.*, 1996
Mungbean	350	26.3	Uprety *et al.*, 1996
	600	39.3	Uprety *et al.*, 1996
Rice	350	17.4	Ziska *et al.*, 1996
	600	22.7	Ziska *et al.*, 1996
Plantago spp.	350	8.3	Stulen *et al.*, 1991
	600	12.7	Stulen *et al.*, 1991
Sweet chestnut	350	6.4	Rouhier *et al.*, 1994
	600	10.1	Rouhier *et al.*, 1994

Water Use Efficiency

At the cellular level elevated CO_2 may cause reduction of the transpiration rate by inducing partial closure of stomatal guard cells on the leaf surface (Jones and Mansfield 1970). This contributes to the increased water use efficiency. Bazzaz and Carlson (1984) found that the interaction of elevated CO_2 and water availability enhanced the water use efficiency (WUE) in the C_3 species than the C_4 species by the ratio of carbon fixed to water transpired. Physiologically WUE increase is one of the most significant plant responses which is identified by the excess CO_2. Elevated CO_2 suppresses the use of water and rise in photosynthetic rate (Uprety *et al.*, 1995), which pushes this important ratio in upward direction. This ratio (Net photosynthesis and transpiration rate) in C_4 species as in corn was 27:73, while in the C_3 species as in soyabean it was 90:10 (Acock and Allen 1985; Rogers *et al.*, 1986b). Water use efficiency increases by CO_2 enrichment has been reported by Baker *et al.*, 1990c; Mosison 1985; Sionit *et al.*, 1984. Data shown in the Table 9.2, and the decrease in rate of transpiration and improving water potential enhances the water use efficiency of the various crops. Allen (1992) indicates that larger plant size (leaf area) counter balances the reduction of water use and offsetting enhances the WUE. This agrees with the hypothesis of Cohen (1970) that the C_4 photosynthetic pathway and higher water use efficiency of C_4 plants do not necessarily yield a competitive advantage under drought conditions. But in C_3 plants water use efficiency is higher with elevated CO_2 and low leaf area index, while the ambient CO_2 and high leaf area index resulted in the lowest WUE (Rogers *et al.*, 1992). Rosenberg *et al.* (1990) has examined the potential effects of CO_2 enrichment on evapo-transpiration with complete simulation which include climate change. He advances the theory of water use efficiency and explains how to decrease transpiration from the plants by restricting the evaporation from the soil surface (Kuchment and Startseva 1991).

Table 9.2: Effect of Elevated CO_2 on the Rate of Transpiration

Crop	CO_2 Concentration	Transpiration ($\mu g\ cm^2 s^{-1}$)	Reference
B. campestris	350	13.9	Uprety et al., 1995
	600	9.2	Uprety et al., 1995
B. carinata	350	11.4	Uprety et al., 1995
	600	7.6	Uprety et al., 1995
Maize	350	2.6	Samarakoon and Gifford 1996
	600	1.5	Samarakoon and Gifford 1996
Tomato	350	6.4	Peaz et al., 1996
	600	3.6	Peaz et al., 1996
Aster	350	21.4	Marks and Strain 1996
	600	13.8	Marks and Strain 1996
Broomsedge	350	8.4	Marks and Strain 1996
	600	3.4	Marks and Strain 1996

Table 9.3: Effect of Elevated CO_2 on the Leaf Water Potential

Crop	CO_2 Concentration	Water Potential (bar)	Reference
B. campestris	350	−5.7	Uprety et al., 1995
	600	−6.0	Uprety et al., 1995
B. carinata	350	−5.0	Uprety et al., 1995
	600	−3.5	Uprety et al., 1995
Tomato	350	−6.3	Peaz et al., 1984
	600	−6.3	Peaz et al., 1984
S. Minor	350	−1.5	Ferris and Taylor 1994
	600	−3.6	Ferris and Taylor 1994
L. corialetus	350	−3.3	Ferris and Taylor 1994
	600	−3.5	Ferris and Taylor 1994
P. mediadge	350	−3.5	Ferris and Taylor 1994
	600	−5.2	Ferris and Taylor 1994

In the comprehensive review of this topic Kimball and Idso (1983) have cited 46 observations which cumulatively show that the high CO_2 reduces the transpiration rate. Samarakoon and Gifford (1986) have found the relative decrease in transpiration rat (29 per cent) due to high CO_2 concentration during the crop period in maize, but no leaf area increases to counterbalance the reduced transpiration per unit leaf area.

Thus, water use per plant remains lower (25 per cent) at high CO_2 throughout the growth period. However, the present study has important implications on modelling of CO_2, crop and global climatic parameters. But all the crop species do not alter or reduce the rate of water use under high CO_2. Plant species with larger and deeper roots are capable of consuming deep water, while those with superficial roots cannot suck water from deep soil resulting in a dry surface or drought. The existence of large number of species with diverse responses in terms of whole plats water use to high CO_2 suggest a possibility of changes in species composition in mixed plant communities with wide range of stomata resistance responses (Easmus and Jarvis 1989; Uprety *et al.*, 1995).

Dry Matter Production

During the past few decades various studies have been conducted on the effect of CO_2 on total dry matter production of plants (Kimball 1983). Results of these experiments have been depicted in Table 9.4. Which show that elevated CO_2 concentration increases the total dry matter both in C_3 plants (Uprety *et al.*, 1996). Root dry matter increases 50 per cent by elevated CO_2 in approximately 50 per cent of the studies containing root data. Roots often exhibit greater dry weight increase among plant organs (*i.e.* stem, leaf) under high CO_2 concentration (Imai *et al.*, 1985; Norby *et al.*, 1992; Rogers *et al.*, 1983a; Wittwer 1978) or an increase in biomass partitioning towards the roots (Hocking and Meyer 1991; Imai and Murata 1976). This biomass partitioning results in an increase in the root to total shoot ratio (R:TS). Increasing the concentration of CO_2 in greenhouse systems increases the percentage of cuttings which forms roots in numerous ornamental and floricultural species (French 1989; Lin and Molnar 1981). Elevated CO_2 during propagation also increases the root number and root length of sweet potato (Bhattacharya *et al.*, 1985).

In addition to the primary root response, other variables *i.e.* root dry weight, root to shoot ratio, root length and root number have positive response under elevated CO_2 levels (Uprety *et al.*, 1996) in the mungbean crop. Similarly, total shoot biomass is also enhanced by 68 per cent but in the case of tree annual crop biomass increases about 20 per cent (Field *et al.*, 1995). However above ground biomass production in the phytocells is increased at least 5 times in the field. It has also been known that resource limitations restrict growth and leaf surface production before intrinsic rate of resources capture *e.g.* leaf photosynthesis per unit leaf area (Natr 1972). As for the photosynthesis and total biomass production, the response is better at elevated CO_2 level (Sinclair and Horie 1989), while the leaf area does not respond proportionately to elevated CO_2 (Baker *et al.*, 1990; Ziska and Teramura 1992). In case of drought, number of leaves and their dry weight decreases but increases at high CO_2 level. In addition, high CO_2 concentration also increases the number of leaf, root growth, total dry weight and number of flower buds in the drought stressed plants. This hypothesis is also supported by Peaz *et al.* (1984). There is also strong evidence that specific leaf area (SLA) decreases with increase in CO_2 concentration. Decreased SLA in high CO_2 grown plants in often associated with increased starch levels in leaves, whereas it increases nitrogen concentration (Bazzaz 1990) which enhances the carbon nitrogen ratio with the excess carbon accumulation in the leaves (Mishra *et al.*, 1996).

Table 9.4: Effect of Elevated CO_2 on the Total Dry Matter Production

Crop	CO_2 Concentration	Dry Matter (g/plant)	Reference
B. campestris	350	25.2	Uprety et al., 1995
	600	28.2	Uprety et al., 1995
B. carinata	350	32.0	Uprety et al., 1995
	600	34.5	Uprety et al., 1995
Mungbean	350	3.2	Uprety et al., 1995
	600	3.5	Uprety et al., 1995
Maize	350	7.6	Uprety et al., 1995
	600	8.0	Uprety et al., 1995
Plantago mayer	350	12.9	Den Hertog et al., 1996
	600	13.1	Den Hertog et al., 1996
Rice	350	6.0	Ziska et al., 1996
	600	8.0	Ziska et al., 1996

Seed Yield

During the last decade there have been extensive researches and reviews on the effects of CO_2 enrichment on photosynthesis, crop growth and yield (Kimball 1989; Lemon 1986; Cure and Acock 1986; Allen 1990).In a survey of 430 research experiments on various crop species, Kimball (1983) reported that an increase in atmospheric CO_2 would produce a higher seed yield about 33.6 per cent especially in C_3 plants. Enriched CO_2 concentration increases the global temperature by arresting the solar radiation. This may enhance the yield due to a decrease in the duration of grain filling (Eamus 1991). Gifford (1988) has also used computer simulation model for prediction of wheat yield and has observed that the increase in yield by enhanced CO_2 concentration can compensate up to 2-3°C increase in global mean temperature. However, he has also suggested that in specific cases where terminal drought occurs at the end of the growing season, the grain filling rate is reduced, but the CO_2 enrichment stimulates the yield, hence the impact of drought is nullified.

Effects of CO_2 on crop growth and development are myriad, but the most important ones could probably be lumped together under the categories of biological nitrogen fixation, photosynthesis, water use efficiency and yield component. Lemon (1977), Wittwer (1980), Allen (1979), Kramer (1981) and Rosenberg (1981) have reviewed these aspects for elevated CO_2 and found overwhelmingly its positive effect on yield. Mean yield of various crops obtained using controlled CO_2 enrichment chamber is depicted in Table 9.5. Significantly, enhance in the mean yield is reported to enrichment of CO_2. Upreaty et al.(1996) has reported 15-30 per cent increase in seed yield by doubling the atmospheric CO_2 especially in mungbean and *Brassica* species. Havalka et al. (1984) and Bhattacharya et al. (1985) have reported that though the number of seeds per pod decreases, the seed weight and number of seed/plant significantly increase at elevated CO_2 levels. In case of Bean and Mungbean plants, high CO_2 concentration increases the seed yield by increasing the number of pods (Gustaftion 1989; Uprety *et al.,* 1996).

Table 9.5: Effect of Elevated CO_2 on the Seed Yield

Crop	CO_2 Concentration	Seed Yield (g/plant)	Reference
B. campestris	350	36.3	Mishra *et al.*, 1996
	600	46.3	Mishra *et al.*, 1996
B. carinata	350	80.5	Mishra *et al.*, 1996
	600	96.9	Mishra *et al.*, 1996
Mungbean	350	22.0	Uprety *et al.*, 1995
	600	33.0	Uprety *et al.*, 1995
Maize	350	45.0	Uprety *et al.*, 1995
	600	47.0	Uprety *et al.*, 1995
Wheat	350	34.1	Kimball and Idso 1992
	600	38.3	Kimball and Idso 1992
Gram	350	24.2	Ziska *et al.*, 1996
	600	27.4	Ziska *et al.*, 1996

10

Environmental Pollution and Plant Health

Definition of Environmental Pollution

The presence of substances in the ambient atmosphere resulting from the activity of man or from natural process causing adverse effects to man and the environment (Weber, 1982). An expanded view of this definition, the presence in the atmosphere of substances or energy in such quantities and of such duration liable to cause harm to human, plant, animal life, damage to human made materials, structures and changes in the weather and climate, or intervene with the comfortable enjoyment of life or property or other human activities.

Photochemical Oxidants

Ozone, an oxidant much stronger than oxygen, causes cracking of stretched rubber at hourly concentrations of only 0.01-0.02 (20-40 $\mu g/m^3$), although ozone inhibitors can be built in to rubber products such as vehicle tyres and rubber insulation, ozone also attacks the cellulose in textiles, reducing the strength of such items and all oxidants cause some fading of fabrics and dyes. Textile fabrics affected include cotton, acetate, nylon and polyester, (Shaver et al., 1983).

Oxidant cause acute and chronic injury to plants, causing necrotic patterns on leaves growth alterations, reduced yields and reductions in the quality of plant products. Typical affected zone are stippling or flecking (brown spots or flecks which subsequently turn white) on the upper surface of leaves. (Miller et al., 1969). The large are of trees are dying as the result of prolonged exposure to photochemical oxidants.

The trees are injured by the high ozone concentration and become very susceptible to fatal pest outbreaks of pine-bark beetles. Nevada Mountains, where ozone levels are much lower, widespread injury by chlorotic mottle on the needles are more advanced stages of chlorotic decline have been attributed to the smog. Many parts of Europe die-back of forests appears to be happening on a massive scale. Air pollutants, including ozone are considered to be largely responsible for the damages to European forest (Hinrichsen, 1986, Skarby and Delrden, 1984).

Carbon Dioxide and Ozone Depletion

The net effect is that some pants should be more resistant to water stress (and more tolerant to atmospheric pollution as partially closed stomata impede entry of potentially harmful air pollutants into leaves) Sionit *et al.* (1981) found that CO_2 fertilization completely compensated for lack of water when wheat was grown in the laboratory without enough water for maximum development. While water stress periods depressed production of wheat grown under normal carbon dioxide levels, wheat grown under 1000 ppm of CO_2 produced as much as unstressed plants grown at the current atmospheric level of CO_2 and remained turgid of moisture levels that wilted their unexposed counter parts. Reverse and Shapero (1978), point out that its average cloudiness the quantity of incoming sloar radiation will be lowered and the energy available to crop plants for photosynthesis will diminish. However whether changes in climate will offset the crop yield grains attributable to direct enhancement of photosynthesis by higher levels of CO_2 will continue to be debated (Waggoner 1984).

Acid Rain

Effect on Terrestrial Ecosystems

Acid deposition may cause damage to terrestrial ecosystems by increasing soil acidity, deceasing nutrient availability, mobilizing toxic metals leaching important soil chemicals and changing species composition and decomposed microorganisms in soils. Much concern has been expressed that acid precipitation causes a reduction in forest productivity. Large areas of forests, especially at altitudes about 600m have been damaged in Europe with significant proportion or even killed (Elsoon 1987). In West Germany, four of the nation's most important tree species, Norway spruce, white fir, scotch pine, and beech have shown alarming signs of deterioration (West Stone and Foster, 1983). Different trees are affected at different rates, on spruce top growth slows resulting in a "stork" "nest" effect. Needless begin to yellow, usually on the upper side of branches, and then fall off. Growth appears stunted and side branches die. In deciduous trees, leaf discolouration, early leaf fall, death of tree tops, damage to bark as well as lack of natural rejuvenation is observed.

Ulrich *et al.* (1980) and Ulrich and Pankrath (1983) showed that acid rain is leaching important plant nutrients such as calcium, magnesium and potassium from the soils, making them unavailable to trees in addition acid rain mobilizes aluminium in forest soils (from harmless soil compounds such as aluminium silicate) which decreases the ratio of calcium to aluminium in soil solutions to the extant the root growth is impaired. The fine root tree are damaged by the toxicity of aluminium, which together with nutrient deficiency, causes a stress condition of crown die-back (leaves or needless at tree top turn yellow, then brown and eventually drop off) and the eventual death of the tree. Seedlings may not survive because of extended damage to their roots.

To the increasing concentrations of photochemical oxidants (CPAN, ozone) in Europe, to increase in oxidants of nitrogen and hydrocarbons from motor vehicles and to synergistic effects between many of these forms of pollution. What ever reason, it is recognized that stressed trees then become vulnerable to further damage in through drought, wind damage (through weakened root systems), snow damage (the brittle tops of trees snap beneath the weight of snow) insect (*i.e.* bark beetle) fungal and virus attack.

Einbender *et al.* (1982). Observed die-back of red spruce in Vermont and New Hampshire. United Nations Economic Commission of Europe (1985) explaining the wide spread damage of forests in Europe is made more difficult by the existence of only limited acid rain effects on the extensive forest of Southern Scandinavia. However acid rain deposition rates in Scandinavia are considerable less than those observed in central and eastern Europe where tree damage such as reduced foliage and brittle crowns may not expected. Negative effects of leaching of nutrients may be temporarily offset in established forests by the increasing fertilizing effect of nitrates. In long term, forests productivity may be limited by the lack of nutrients other than nitrogen. To compensate for decreasing soil mineral reserves, the trees overall root mat system eventually shrinks, making it susceptible to drought in summer and wind throw in winter ever fertilization of nitrogen may also makes trees to more palatable for insects. Gorham (1982) suggest that the acidity may hinder the ability of the forests bacteria and fungi to recycle nitrogen from decaying plants into the soil, robbing the trees of such nitrogen as the precipitation deposits.

Record *et al.* (1982) report on an thirty two commercial crops grown under controlled environmental condition and exposed to simulated acid rain of pH 3.0, 3.5, 4.0 and 5.7 preliminary results indicate that same crops suffered extensive foliage damage and yield reduction but others sustained little apparent injury even under these severe exposure conditions. Similar variability also reported by Hibbard (1983) when twenty seven crops tested, five were inhibited, six stimulated and sixteen experienced no effect of acid rain.

Waste Heat

Studies by Changnon (1977) impact of waste heat on urban modification on summer weather in and around St. Lovis suggest that the resulting changes produce greater disadvantages than benefits for the community. Although increased rainfall leads to increased crops yields. This is more than offset by increased flooding, soil, erosion and acid rain, together with increased hail damage to crop and property.

Oxides of Nitrogen

Effect on Plant Vegetation

Oxides of nitrogen rank second to sulphur compounds in their contribution to acid rain which may affect terrestrial and aquatic ecosystems. United States for example, 30 per cent of the acidity of precipitation (below pH 5.6) is believed to be caused by the oxide of nitrogen producing nitric acid, 65 per cent to sulphuric acid and 5 per cent to hydrochloric acid. However, where as the contribution of sulphate to the problem of acid precipitation is leveling off. That of nitrate is increasing (Babich *et al.*, 1980). Indirect effects of oxides of nitrogen via acid precipitation on vegetation, prolonged exposure to nitrogen dioxide concentrations of 470-1880 µ g/m³ (0.25-1.0 ppm) may suppresses growth of such plants as tomato, pinto beans and navel oranges (Wark and Warner, 1981).

Pollutants and their Effects

But technological improvement alone can not solve the environmental pollution problems because acute air pollution problems may be solved by converting the air pollutants into a solid or sludge this results contamination of land and water, in long terms this waste materials goes to environment anyways. Therefore this problems tackled by the holistic and preventive manner. A holistic approach to solved pollution problems through the best available option have taking into account technical and economic considerations. After the few years, the only way to reduce the pollution problem is to reduce generation do all wasters. Energy, transportation, industry, agriculture and land use management policies need to give much greater consideration to anticipating and preventing pollution problems. Curative pollution control policies or no substitute for preventive policies (Derek Alson-1987).

Environment can be considering fewer than two ways – physical environment and social environment. Physical environment (Biotic) are of major concern for environmentalist today- our unplanned developmental efforts, blind us towards material progress and narrow self interest has lead us to the dark age. The present state of pollution has a great impact on our daily life. All living beings on the earth are breathing polluted air, polluted drinking water, and eating polluted food. Even the underground water and upper and lower layers of the earth contain highly toxic materials adversely affecting plant and human body, mind and behavior. Man started cleaning more and more land, removed diversified vegetation and replaced with selected species of crop plants for grains, vegetables and fruits. In this process he had to face a stiff competition with his rivals. The pests and pathogens soon it has been realized that this is going to be a losing game to man, he did not care whether it was a renewable or non-renewable resource. The use has almost become indiscriminate and uncontrollable. Man has realized that in spite of all ingenuity and resource utilization he can not escape starvation, if he continues to proliferate at the present rate, he has been enough damage to the environments in several ways-resource mismanagement, mitigation of animal and plant wealth of the planet and feuling the atmosphere, hydrosphere and biosphere. We can not bring back those animal and plant species that have become extinct, we can take measure to conserve what is left *i.e.*, wild life, the fauna and flora, forest, non-renewable resources and minimize the usage of resources so that consumption is equaled by renewal because we can not avoid the use of fossil fuel like petrol and coal. We can not ask to former for the use of fertilizer and pesticide. These things are necessary for development. The solution is only for start to think about advance technology adaptation which is environmentally sustainable.

Mishra *et al.* (1999) reported the response of *Alacia nilotica, Terminalia arjuna* and *Casuarina equisetifalia* is fairly better towards effluent water of industrial waste. The survival percentage and growth parameters of different species was satisfactory and recommended for a forestation of site specific tree species play a vital role in abatement of environmental pollution by their ameliorating effects vis-à-vis in increasing the productivity in term of fuel, fodder and timber.

Energy

Increasing consumption of energy to meet the demands of civilization has led to many environmental issue *e.g.* air and water pollution, spread of toxic materials. Scarring of land by strip mining etc. (Abelson and Hamm and -1978) it is commonly accepted that the standard of living, if increase then the consumpton of energy automatically increased but the energy can not fulfill our demand to longer period. Therefore efforts should be made to improve the efficiencies of power plants and energy should be save due to locomotive losses. In addition to, fossils fuel resources are dwindling over the years and are likely to be exhausted in the next century. The only alternative is left to us is to harness solar energy which is be the renewable and non-polluting and providers an ideal energy source.

Deforestation

Deforestation in the hills causes avalanches flooding in the plains during monsoon. It is also responsible for addition of 33 per cent of CO_2 to atmosphere which causes global warming. Forests are demand to be global sinks for CO_2 so that loss of forest cover raises CO_2 levels in the atmosphere. It is need for today to restore to reforestation and forest management, but we can not maintain at the rate of deforestation to reforestation. It is lost solution to evolve the genetic engineering aquaculture and symbiosis holds the promise for the future.

Pesticides and Fertilizers

Man intend towards the intensive agricultural practices for more food production with application of more fertilizer, weedicide, insecticide and fungicides. But he did not realize the future consequences of these practices. They are himself injecting poisons in the environment. Fertilizers and pesticides, applied to crops are largely retained by the soil (Elwards 1973). They become part of environmental cycles by absorption by soil, leaching into water, pesticide degraded and goes to soil and photochemical reaction take place pesticidal residues in crops and feed products also cause long term health hazards. The environmental health aspect seek a solution by applying new biotechnology in place of fertilizers and pesticides *e.g.* biological agents. Environmental audit is undoubtedly the one area of environmental awareness that any organization ignores at its peril. It is also massive, complex amorphous and rapidly developing area. Environmental audit touches many area of current proposed and possibly future environmental regulation and serves, for all organizations as a first step towards environmental sensitivity and as regular and essential part of environmental management systems.

Nevertheless, if business their required all its kills at planning and innovation, it is surely in the area of environmental use will never face more difficult problem than to ensure that our products, factories. Process and distribution system are not only unclaimed aging to the environment, but in same way or other even enhance it. That is the standard to which our children and grand children will hold us.

Popularization of Environment Management Sensitivity

Human are living in the environment and depend upon environmental resources for all necessary ingredients. These ingredient obtained from the natural resources by insensitive manner which are not remain in future. Because all manner of deteoration of natural resources cause many problems related to base of life *viz.* scarcity of safe water, air, soil and vegetations. Developed countries utilized maximum resources for our economic development, where environment are not suitable for human because he did industrialized in the cast of environmental detioration. These countries never seen any boundary for distruction of environmental resources and utilized high technology. Such technology was not sustainable for environment. It is lesson for the people that if the biological environment had been lost the live can not save in the earth. Therefore its need to popularization of environment management strategy for saving natural resources and utilization of resources with sustainable ways such as:

1. Grow more trees and save each tree.
2. Avoid the use of pesticide if it is need utilized sustainable levels.
3. Conservation of biodiversity.
4. Active participation for conservation of forest and wildlife.
5. Avoid the spread of house waste on the road collect and safety places.
6. Avoid the utilization of soap when you are bathing in ponds water, Reservoir Lake and rivers.
7. Not throw waste (Latering, urine) in ponds and rivers.
8. Treated industrial sewage and sludge should be release in the rivers.
9. Do not throw unburned dead bodies in the rivers.
10. Stop the utilization of polythene bags.
11. Treat the industrial gases for safe release in the atmosphere. It should be release through chimney.
12. Regular checking of automobile for saves release of exhaust.
13. Minimum use of loudspeaker and horn.
14. Conserve rain water more and more.
15. Build new ponds and clean old ponds or water reservoir.
16. Use product which is made by without CFC.

Biotechnology as Environmental Management

Demands of more food for increasing population are major cause of shrinking the forest area. Because the productivity of the yield per unit area are not much increasing as per need. Currently the world population is around 6 billion and it is expected to be more than 9 billion by 2050 (Madhyastha 2003). This increasing growth further put pressure on production of food and contribute to convert forest area for cropping. Besides to this, other developmental work and sheltor will be build up on forest land. Therefore to meet the future and present food demands through the crop plants as need to improved technology beside to heavy use of insecticides, fungicides, herbicides and chemical fertilizers because overuse and unscientific use of these chemical further create environmental problems. Over ruled this problems by making it possible to grow more food with added nutrition on less land and under tough situation. Biotechnology is becoming an important part to ensure food security without deterioration of environment. The biotechnological tools are here discussed are not exhaustive but have direct impact on increasing food production with safe environment.

Transgenic Plant Technology

1. Reduced agrochemical use
2. Self protecting plants
3. Herbicide and virus resistant plants
4. Overcome adverse soil conditions
5. Heat and cold stress tolerance
6. Nutritious food

A major goal of plant genetic engineering is the introduction of agronomically desirable phenotypic traits into crop plants in situations where environmentally sustainable methods have been unsuccessful. Targeted insect pests become resistant against insecticide and fungicide, in view of the estimated production losses world wide and the heavy costs of protective treatments, is very important (Vaughan *et al.*, 1987). Transgenic plant technology introduce foreign genes into many crop plants, which are of agricultural importance include herbicide resistant and virus resistance, attempts to be made insect resistance using the *Bacillus thuringiensis* endotoxin genes which have been introduced tobacco and tomato (Falck 1987, Fishoff 1987). Genetic engineering of crops offers opportunities to establish multi-line mixtures by creating new isogenicline containing different resistance genes. Such multi-line however are likely to be effective since sowing intimate mixtures of contrasting varieties in the same experimental plots has led to slower build-up disease (Wolfe and Barrett, 1980). Progeny of transgenic tobacco plants expressing the cucumber mosaic virus (CMV) coat protein (CP) or its antisense transcript (antisense CP) were protected from CMV infection, symptom development and virus accumulation were reduced or absent in CP plants, independent of the strength of the inoculums. These results demonstrated that it is possible to obtain protection in transgenic plants using antisense CP message (Cuozzo *et al.*, 1988).

Transgenic tobacco plants expressing the CMV-D coat protein were protected from infection by CMV-C on both inoculated and systemic leaves. The number of plant that showed lesions were reduced and the symptom development on the inoculated and systemic leaves was delayed compared to controls. Similar results were reported for transgenic plants expressing CP genes for TMV (Abel *et al.*, 1986), AIMV (Tumer *et al.*, 1987) and PVX (Hemenway *et al.*, 1988) potatoes are also affected by a number of bacteria such as Erwinia species, causing soft-rot of potatoes and Pseudomonas causing bacterial wilt. The giant silk moth, Hylophora cecvopia contain same bactericidal proteins. The gene coding for these proteins have been introduced in the potato genome and it is hoped that the transgenic plants would be resistant to these bacteria.

Plant Nutrients

Nitrogen is the most important nutrient for crop production. While some of the plants like pulses and other legumes have the capacity to fix atmospheric nitrogen with the help of bacteria, living in the nodules in their roots, most others depend for the nutrient on the application of fertilizer and manures. Because of economic limitation most of the farmers do not apply the fertilizers and reduce the yield of crop plants. Chinese scientists have genetically engineered the bacteria *Rhizobium spp.* The genetically altered bacteria are now spread over an area of 10 square kilometers in China and enable the farmer's to harvest higher levels of rice, soybean and wheat.

Overcoming Adverse Soil Condition

Soil toxicity affects a very large percentage of arable land and limiting the use of cultivation due to acidity of the soil. This problem has been overcome by use of transgenic technology which able to produce about ten times more citric acid than normal plants and this helps them to grow and produce normally in these acidic soils. Statement of glycine betanin gene from *Escheriaha coli* bacterium helps to tolerate salt concentration three times more than a normal plant can tolerate (Madhyastha 2003). This method create opportunities to grow more forest for sustaining environment and grow more crops for food security on unutilized soils, which were not suitable for plant growth.

Microbial Degradation of Toxic Pollutants

The mixed culture of microbes mixed with unexpected metabolic abilities to degrade recalcitrant environmental toxic waste. These organisms can be further endowed with additional capabilities through genetic engineering and used to clean up toxic wastes such as dioxins, polychlorinated biphenyls, chlorinated phenols and chlorinated benzenes. Mixed cultures are often efficient, but they may not use the toxic chemical as their primary energy sources. In such instances a second compound is used as the energy source and the toxic compound is slowly degraded in a fortuitous reaction called co-metabolism novel microbial pathways slowly evolve to metabolize environmental chemicals if the section pressure is low. The diverse catabolic reactions scattered among various microbes can be marshaled in one organism by the methods of recombinant DNA technology to metabolize chemicals resistant to biodegradation Ramos *et al.* (1987) are trying to broaden the pathway in *Pseuldomonas putida* which degrades methylbenzoate and 3-ethylbenzoate but not 4-ethylbenzoate. The protein that stimulates synthesis of methylbenzoate metabolizing enzymes in *P putida* does not recognize 4-ethylbenzoate. Further more, an intermediate of 4-ethylbenzoate metabolism is suicide substrate and kill the enzyme catechol 2, 3-dioxygenase which is involved in the metabolism of 4-ethylbenzoate. *AP. putidia* that has the mutant regulatory protein to recognize 4-ethylbenzoate and engineered catabolic catechol 2, 3-dioxygenaze that is not killed by the suicide intermediate of the catabolic pathway have been constructed to degrade 4-ethylbenzoate. In the cases of agricultural surpluses and abundance of organic waste material, large-scale biotechnology may be involved. Sugarcane is used for producing more than million tons of ethanol annually. Increasing petroleum fermentated alcohol are very attractive as a fuel. Genetically engineered ethanol tolerant yeast that could produce high yields of ethanol and subsequently serve as cattle feed could be promising (D, Amore and Stewart 1987).

Ecology and Epidemiology

Molecular biology provides useful foods for studying pathogen ecology or epidemiology for sustainable protection. DNA probes can serve to identify and to differentiate pathogen isolates (Leach and White 1989). Before DNA probes were developed. Isolates of the rice blast fungus *Pyricularia oryzae* from used and rice hosts can only be differentiated as the basis of host range. This being utilized for whether population of the fungus are important sources for epidemics rot (Leung *et al.*, 1989). This determination employed for use of fungicides to control epidemics with the successful construction of engineered microorganisms and plants came environmental issues and concerns. The use of such materials in pest management will necessarily require the release of such materials even into new environments. Although there is no consensus yet as to what constitutes a safe release, genetically engineered organisms may play a major role in the environmental management strategies of future (Lindow *et al.*, 1989).

Herbicide Resistant Transgenic Crops

Some herbicides are effective, non toxic to animals and are rapidly degraded in the soil. These environmentally safe herbicides are however non-selective and thus can be used only before emergence of the crop. Through genetic engineering herbicide resistant plants can be produced, expanding utilization of the biodegrade, broad spectrum herbicides (Shah *et al.*, 1986). Biolophos and glutoseinate are examples of such herbicides. Transgenic plants tolerant to glyphosate another non-selective herbicide have also been produced (Fillati *et al.*, 1987).

Conservation of Plant Genetic Resources

The process of genetic erosion necessitates measures that Germplasm must be conserved in such a manner that there are minimal losses or changes in genetic variability of the population. Once a particular type of germplasm is collected from different parts of the world, its maintenance at the breeding station is also necessary because every times it can not be procured. The conservation by the conventional methods of germplasm are prone to possible catastrophic losses due to (1) attack by pest and pathogens (2) climatic disorders (3) natural disasters (4) political and economic causes. For avoidance of such limitations, a suitable tissue culture propagation and conservation of technology should be practised.

Two basic approaches are used to maintain germplasm *in vitro* (1) minimal growth (2) cryopreservation. Minimal growth conditions per short-term storage which can be followed in several ways reduced temperature and light, incorporation of sublethal levels of growth retardants, induction of osmatic stress with sucrose or manital, and maintenance of cultures of a reduced nutritional status particularly reduced carbon, reduction of gas pressure over the cultures, discontinued mineral are overlay. The advantage of this approach is that cultures can be readily brought back to normal cultural conditions to produce plants on demand.

Cryopreservation at ultra-flow temperature of liquid nitrogen (196°C) offers the possibility for long-term storage with maximum phenotypic and genotypic stability. This method is relatively convenient, environmentally sound and economically feasible for large number of genotypes production.

Insect Resistant Transgenic Plant

Sustainable developments of insecticides are needed for future investigation. For this purpose B.t. toxin have been investigated for safe and effective insecticidal properties. Alternative approach to utilizing the B.t. toxin is to isolate the gene, place it under the control of regulatory sequences that function in plants and transfer the gene into the plant genetic materials. The B.t. gene has been isolated, incorporated into agrobacterium T-DNA and used to transfer tobacco (Vaeck 1987, Hilden *et al.*, 1987). Insect tolerant transgenic tomato plants have also been produced using the same strategy (Fischh off *et al.*, 1987).

Table 10.1: Indian Air Quality Standards ($\mu g/m^3$)

Sl.No.	Statistics	NO_2	SO_2	SPM
1.	Industrial area	120	120	500
2.	Residential area	80	80	200
3.	Sensitive	30	30	100

Source: CACB, 1996.

CPCB Report (1996) Ambient air quality status and statistics CPCB publications, Delhi pp. 1-25.

The success of environmental protection depends upon women's empowerment to negotiate and execute the daily activities for which capacity building to frame their life style leading to enhancement of quality of life of the five capital assets: human, social, physical, financial and natural. The first one is in plenty and the lost is dwindling fast because of which the survival of the farmers is at stake. (Kantkar and Mistry 2000). Indian women share the wisdom they have with family and society and employed the same for conservation work. They have evolved as good caretaker of the environment as they have been worshipping mother earth through the ages with great pomp and pride.

11

Source of Environmental Pollutant and Plant Health

C lean air and water are an aesthetic pleasure, when life of city have all but forgotten the smell of fresh air and clear glittering lakes, streams and rivers. Cultural memorial tomb like Taj Mahal, the Parthenon in Athens and statue of liberty in New York are in increasing danger of being destroyed by the atmospheric pollution. This resulted into unlimited exploitation of every bit of natural resources. The natural resources are utilized for rapid industrialization of the countries. These industrial products are making our life comfortable but the unfavourable atmospheric conditions created by man himself threatened the survival not only of man but also of other living organisms. The various by products generated by the industry and subsequently thrown in their waste are a matter of great concern because these by- products are not only toxic but concerous and recalcitrant as well. The serious attention of human towards the dangers caused by environmental pollution come in 1952 when nearly 4000 people died in London, U.K. because of "Killer Smog" that settled all over the city.

The irresponsible activities of man had adverse effect on living organisms in the biosphere. The planet earth including atmosphere (air, land and water) that sustain life is called the biosphere. Biosphere extends up to about 7 km of the earth surface itself, downward into the ocean to depth of about 10.67 km and vertically into the air to about 10 km where the important events, such as cloud formation, lightning, thunder storm formation etc. all take place in this region. These events provide splendid nature for life be exist. The industrial infrastructure like metals, chemicals fertilizers, petroleum, food and its products such as pesticides, detergents, plastic, solvents, fuels, paint dyes, food additives etc are some of major pollutants.

The tremendous increase of atomic energy also increased radioactivity in the atmosphere. Exhaust emission of oxides of nitrogen, carbon monoxides, hydrocarbon, sulphur dioxide and lead added poisonous gases in the atmosphere of urban areas. The emissions of nitrogen oxides and hydrocarbon also produced the secondary pollutants of photochemical oxidants.

Pollution is a necessary sin of all developing countries for development, but the lack of pollution control systems in the industries create heavy backlog of pollutant in the biosphere which causes sickness in the crop plants. Until the 1960's pollutants were generally regarded as a problem in the vicinity of individual emission source or within urban areas. Now-a-days pollutants are transported over long distances and causing adverse effects on the environment at the long distance or rural areas. Long-range transport of sulphur and nitrogen compounds across national boundaries resulted in increased acidity of precipitation (acid rain) at distant locations and created International problems.

What are Pollutants?

Any substance which causes pollution is called pollutant. A pollutant may include any chemical or geochemical (dust, sediment, arid sand, pollen etc.) substance biotic components or its product or physical factor (heat, cold, snow, steam, fogs etc.) that is, released intentionally by man into the environment. Some pollutants which are man made magic compounds like DDT, which were thought to be harmless to humans (Paul Mullar was awarded Nobel Prize in 1958 for high contribution for controlling Malaria using DDT) have now been recognized as a cumulative toxin for humans. Ponds, lakes and rivers are polluted by waste strain chemical, heavy metals-pesticide, fertilizers and other factors and air by gases of automobile exhausts, industries thermal power plants etc. Incidents such as the tragedy at Bhopal, India in December, 1984 in which 2500 people died when Methyl isocyanite was accidentally released from the union carbide chemical plants. (The 1972 United Nation Conference on the Human Environmental at Stockholm). The late Prime Minister of the India Smt. Indira Gandhi had said a Modern man must re-establish an unbroken link with nature and with life.

Environmental Pollutants

Environmental pollutants which are unfit for use of air, water and lands are as given below.

1. Deposited matter: Smoke, dust, grit soot, tar dyes, paint etc.
2. Gases: Oxide of nitrogen (NO, NO_2^{---}) SO_2 sulphur dioxides, carbon monoxide, carbon dioxide, chloroflurocarbon, halogen, chlorine, bromine, iodine.
3. Acid droplets: Sulphuric acid, Hydrochloric acid, nitric acid.
4. Fluoride

5. Metals: Cadmium, nicklel, cobalt, gold, silver, copper, manganese, zink, tin, mercury, lead, iron, chromium, uranium.

6. Agrochemicals: Pesticides, herbicides, fungicides, nematicides bactericides, fertilizers etc.

7. Organic substances: Benzene, ether, acetic acid benzopyrences.

8. Photochemical oxidants: Photochemical smog ozone, peroxyacety nitrate (PAN), peroxy-benzoil nitrate (PB_2N), nitrogen oxides, aldehydes, ethylene.

9. Solids wastes.

10. Radio active waste.

11. Noise: Jet, engine, rivet gun, propeller, air craft.

12. Odours: Car, bus, Heavy lorry, Loudspeaker etc.

13. Waste Heat.

Types of Pollution

Pollution is divided on the basis of environment pollution. These pollutant goes into environment in the following ways.

1. Air Pollution
2. Water Pollution
3. Soil Pollution
4. Marine Pollution

There are various varieties of pollutants in the environment but we have classified them into two basic types –

1. Biodegradable pollutants
2. Non-biodegradable pollutants

1. Biodegradable Pollutants

Domestic waste that can be rapidly degraded under natural condition such as vegetable wastes, animal dungs of cows, buffaloes, horse, sheep and goats. Solid sludge of agricultural based industry which is easily degradable like sugarcane industry, cotton industry, tomato, potato and aonla based industry. Their waste create pollution problems when they accumulate in the open space. Transformation of organic compound by microorganism, consisting of large molecular (proteins, long-chain hydrocarbons) carbohydrates fats into smaller sized molecular compounds and eventually into simple inorganic substances such as CO_2, H_2O and NH_3. The decomposition can be either aerobic or anerobic and play an important role in the biogeochemical cycle and in the self purification of water.

2. Non-biodegradable Pollutants

Metal based industries produces many poisonous substances which are either not degradable or very slow degradable like mercuric salts, aluminum cans, phenolic compound, lead DDT etc. They are not cycled in biosphere in spite of their non-degradable nature. They not only accumulate in the atmosphere but are often biologically enlarged with their subsequent movement in food chains. The accumulation of persistent organic substances means another source of danger for the environment, from this the quantity of biologically nondegradable such as construction rubble, plastic, gas, tins and car is also increasing.

Air Pollution

Detrimental changes in the normal composition of air are affected by air pollutant *e.g.* dust, aeosals of different composition, sulphur dioxide, carbon monoxide, nitrogen oxides etc. Air pollutants originate from industrial and domestic heating, factories, vehicles etc. The earth's vertically extended atmosphere, an envelope of gases is divided into the following layers.

  i. **Troposphere:** The lowest atmosphere upto 5 km in which temperature decreases with height bounded by land or sea surface below.

  ii. **Stratosphere:** The region above 5 to 45 km where temperature increases upto 90°C with height.

  iii. **Mesosphere:** The part between stratosphere and thermosphere from 45 to 80 km. The temperature again decreases upto-80°C.

  iv. **Thermosphere:** Above 80 km the upper part in which temperature increases with height. There is boundary between the atmosphere.

About 75 per cent of the earth's atmosphere belong within 16 km of the surface and 99 per cent of the atmosphere lies below an altitude of 30 km.

The atmosphere has a capacity for dispersion and dilution, it is used as if it is a gaint channel (Munn and Bolin, 1971). Atmosphere is an insulating blanket around the earth. The amount of pollutants introduced into the air through this channel has proved to be restricted and its diluting capacity. It is a source of essential gases, maintains a narrow difference of day and night temperatures and provides a medium for long-distance radio communication. It also acts as shield around the earth against lelhal UV variations and meteors. Without atmosphere, there will be no light, no wind, no cloud, no rains, no snow and no fire. Recent studies have called attention to the harm arising from air pollution, which threat both the physical and chemical conditions of the whole atmosphere and indirectly the biological equilibrium of the entire Earth.

Pure Air

Atmosphere air belongs to the group of incoherent, coarse dispersal systems, it is a system of colloidal particles dispersed in gas, in which solid and liquid components can be found in a mixture of gases. Actually pure air never existed in the atmosphere. Stern (1962-68) describes the pure air is composed of various amounts of substances additional to its constant components. These additional components form the objects of studies on air pollution such as cosmic dust, dust driving from land surfaces, combustion products of forest fires, gases of volcanic eruptions have existed from the very beginning. Air borne microorganism pollen and spores, *i.e.* the multiplicity of floating organisms can also be grouped into this category. The concept of pure air can be defined from the view point of biology and hygiene, unless the amount of air pollutants does not exceed the determined limit. Such air is called pure.

Table 11.1: Gaseous Composition of Unpolluted Air (Dry basis)

Constituent	parts per million (ppm) by Volume
Nitrogen	780,900
Oxygen	209,400
Water	-
Argon	9,340
Carbon dioxide	330
Neon	18
Helium	5.3
Krypton	1.0
Nitrogen oxides	0.5
Hydrogen	0.5
Xenon	0.08
Organic vapours	0.02

Primary sources of pollutions produced and introduce contaminants into a air. As results of primary sources of pollutant activity is called secondary source of pollutants.

SOURCE OF POLLUTANTS

A. Primary Source

1. Natural Origin

A large number of air pollutants is of natural origin. The hydrosphere produces a significant quantity of aerosols (Dispersed solid (smoke) or liquid (mist) particles of colloidal size in the air). Drops of water which get into the air in the course of wave motion of the sea, evaporate and producing Na, Ca, K, chloride, sulphate ions of these. Marine plants and animals produce a large volume of gaseous materials. These depending on partial pressure and temperature, which dissolved in water and introduce in atmospheric carbon dioxide is the most important of these.

Solid state pollutants originate mainly from the surface of lithosphere, desert and sea-sand (SiO_2) depending on which frequently occur in the atmosphere. Soil dust also contains organic components. Though it is composed mainly of the dust of minerals, carbonate, sulphates and oxides ($CaCO_3$, $CaSO_4$, $MgCO_3$, Al_2O_3, Zno, Sio_2). Volcanic activity also introduces dust fumes and gases (H_2S, SO_2, HCl, CO, CO_2) into the air. Decomposed products of plant and animal origin, especially those containing sulphur have frequently an offensive odour (ammonia, hydrogen sulphide and mercaptons, indole and skatols etc). Vegetation primarily and fauna secondarily affect the natural balance of oxygen and carbon in the atmosphere. Hydrocarbons of small atomic number are emitted into the air either by discharges of gas or in the form of organic decomposition products. Methane is a component of the carbon cycle, it can be detected in low concentration from the vegetation floating microorganisms of the air can also be detected such as viruses, bacteria and fungi algae and pollens.

2. Agricultural Activities

The use of fertilizers thus mentioned the source of air pollutants with the increased use of chemicals in the agriculture numerous new substances are introduced in air. All these may exert harmful effects on the natural equilibrium. Use of insecticides, pesticides, herbicides, fungicides and nematicides contain chlorinated hydrocarbons, organic phosphate-esters, thiocarbonates or harmones, which are biologically active substances. They are frequently introduced into the air, in the form of dust or spray especially by air craft used in large-scale agricultural production. It has become one of the most threatening source of pollution and endangering biological cycle.

3. Industrial Sources

Industrial developments are the source of air pollution, (Petroleum refineries, cement factories), fertilizer factories are the major source of chief gaseous pollutants like SO_2 and NO_x. These gases are posing threat to Taj Mahal in Agra and other historical monuments like Fatehpur Sikri, Delhi and Hyderabad. Cement factories emit plenty of dust which create health hazard, stone crushers and hot mix plants also create a menace. There are many industries which emit acid vapours in the air *i.e.* food, fertilizer and chemical factories. During last decade a number of industrial unit established in Delhi.

Table 11.2: Main Pollutants

Solid phase	
Dust	Cement dust
Scoot Fly-ash	Textile dust
Mineral dust	Organic dust
Metallic dust	Toxic dust
Liquid phase (Smog, spray)	
H_2SO_3	HCl
H_2SO_4	Cn Hm
Gaseous and vapour phase	
SO_2	Cn Hm polycyclic and aromatic hydrocarbons
H_2S	Cl
CS_2	HCl
CO	HF
CO_2	F_2
NO	O_3
NO_2	Metallic fumes
M_xO_y	Solvent vapours

Central Pollution Control Board made case study of Najafgarh Road, Lawrence road, Wazirpur, Kirti Nagar, DLF industrial area and Motinagar area which has chemical, fertilizer, iron and steel rolling units was found the most polluted area. It has 32 air polluting units which emit every month 75.3 tonnes of SO_2 and 794.5

tonnes SPM. The Lawrence road area has 27 air polluting units emitting 20.4 tonnes of SO_2 and 140.2 tonnes of dust in a month. The Wazirpur area has 90 air polluting units emitting every month 18.2 tonnes of SO_2 and 254.7 tonnes of dust. The Kirtinagar, DLF and Moti Nagar complexes close to one another have 14,15 and 3 air polluting units respectively (Sharma, 1994). The most important and most frequently occurring pollutants grouped according to their physical state are summarized in Table 11.2.

Thermal Power Source

Coal based thermal power plants are chief pollutants of fly-ash, SO_2 and other gases and hydrocarbons. There are a number of thermal power plants in India like-Singrauli in U.P., Korba in M.P., Ramagundam in A.P., Farak in W.B., Indraprastha estate, Rajghat and Badarpur in Delhi. The pollutant gases produced by thermal power plant are given in Table 11.3.

Table 11.3: Gaseous Pollutants of Thermal Power Plants

Carbon monoxide	0.35
Hydrocarbons	0.14
NO_x	14.00
SO_2	13.30
Aldehydes	0.0035
SPM	369.60
Ash content	92.40

Source: Chaudhuri, 1982.

Automobile

Vehicular traffic introduces dust, soot, carbon dioxide and their derivatives, sulphur dioxide, carbon oxides into the air, from among the combustion engines, diesel engines emit, exhaust gas into the air which owing to the combustion taking place at high pressure is rich in nitrogen oxides. The sulphur content of diesel oil reappears in the farm of SO_2 in exhaust gases.

Table 11.4: Solid Pollutant Released from Indian Atomic Plants

	1990	2000
Capacity of pollutant (MW)	800	1000
Solid Waste (m³)	1700	107000
Low grade (m³)	2500	77100
Medium grade (m³)	650	19900
High grade (m³)	350	8000

Source: Civil report.

Exhaust gases of petrol engines contain carbon monoxide, petrol vapour, aldehydes, straight chained and polycyclic hydrocarbons and lead, which is derived from the lead containing tetraethyl additive. The vicinity of air parts increasing number of air polluting effects of aeroplanes, these equipped with turbo and get engines are intensified are two wheeler operating with petrol.

Table 11.5: Contribution of Air Pollution by Automobiles

State	Smoke Intensity per cent
U.P.	50
Haryana	35
Punjab	25
Delhi	50

In major meteropoliton cities vehicular exhaust accounts for 70 per cent of all CO, 50 per cent of hydrocarbons, 30-40 per cent of all oxides and 30 per cent of SPM.

Principal Air Pollutants and their Effect on Plant Health

Table 11.6: Source of Air Pollutants in India

Source	Quantity of Consumption/ Year (Lack ton)	Total Quantity of Pollution (ton)
Coal	675	550
Diesel	59	290
Petrol	16	840
Gas	125	67
Wood	1000	8360
Animal dung	550	1910
Grasses	320	2660
Domestic waste	2000	6200
Total	4750	26030

Source: Environment and Ecology, 1991.

Carbon Monoxide (CO)

Carbon monoxide is a colourless, odourless, gas produced by the incomplete combustion of carbon containing fuels like stoves, furnaces, open fires, forests and bush fires, burning coal mines, factories, power plants etc. These pollutants are exhausted products from motor vehicles in common busy routes and intercrossings in cities such as Delhi, Culcutta, Mumbai, Chennai etc. Every year human activities may put some 1500 tera grams of carbon monoxide into the Earth's atmosphere compared with the 1200 T g/year from natural sources (US National Research Council, 1997).

The major source of carbon monoxide emissions at breathing level outdoors is exhaust of petrol powered motor vehicles, the diesel engine, high concentrations of CO may occur near industrial plants *viz.*, power station, petroleum refineries, iron foundries, steel mills, hot mix plant, stone crushers etc. In air its concentration is from traces to 0.5 ppm, CO levels in urban areas range from 5 to 50 ppm. Natural sources of this gas are various plants and animals. Most plants are not affected by CO levels known to affect man. At higher levels (100 to 10,000 ppm), the gas affects leaf drop, leaf curling, reduction in leaf size, premature aging etc. it inhibits cellular respiration in plants. Gaseous pollutants penetrate into the intercellular tissues via the stomata. They can be fixed on the surface of cells, enter into reactions with water or take part in metabolism. On the other hand gases may also enter into reaction with chlorophyll, thus inhibiting photosynthesis. The epidermis of the leaf is most resistant to these impacts. However, spongy paranchyma and columnar cells, the main places of assimilation are very sensitive. Cells are capable of decomposing and neutralizing low concentrations of pollutants. Beyond a certain limit, however, loss of water, plasmolyces occurs followed by death. More severe damage results in visible injuries. The tissues of the leaf will shrink, shrived and wilt in places.

Carbon Dioxide (CO_2)

Carbon dioxide is a colourless trace gas which occurs naturally in the atmosphere. Major amount of carbon dioxide is released in the atmosphere from burning of fossil fuel (coal, air etc.), from domestic cooking, heating and the fuel consumed in furnaces of power plants industries, hot mix plants. From fossil fuels more than 18×10^{12} tonnes of CO_2 is being released into atmosphere each year. Burning of carbon based fossil fuels and changes in land-use practices, are increasing the global atmospheric carbon dioxide concentration such that this will result in significant increase in surface air temperature and change in other climatic parameters. Many researchers predict the atmospheric CO_2 concentration will increase from pre-industrial value of 260-300 ppm to 600 ppm during this century. Today it is 350 ppm and by the year 2035 and 2040 it is expected to reach 450 ppm. Such increase of CO_2 concentration estimated to cause an increase in the mean global surface air temperature of 3 ± 1.5 °C (USEPA 1983a).

To some extent, an increase in CO_2 level in atmosphere increases the photosynthesis rate and consequently plant growth acting as fertilizer especially in hot tropical climate. This potential of fertilizer effect may be exploited by using modified crop varieties and agricultural practices. However, an increase in CO_2 concentration in atmosphere may result into disastrous effects also.

At full outdoor light intensity, net photosynthesis in many plants increases with increasing CO_2 concentration upto 900 ppm (Cooper 1982). The enhancement depends upon plant biochemistry, growth stage, status of water availability and nutrients in the soil. The plant group of C_3 and C_4 differ in the biochemical pathways which affect the fixations of carbon dioxide in photosynthesis and differ in the degree of benefit from increased carbon dioxide concentration. Ambient levels of carbon dioxide concentration C_4 plants have substantially higher rates of photosynthesis than C_3 plants at all temperature. While C_3 plants have capacity to increase photosynthesis with increasing CO_2 concentration upto 1000 ppm, C_3 plants includes temperate species like small grain, legumes, grasses, forest trees. While C_4 plant species are generally adapted to warm climates and include tropical grasses, corn, sorghum, maize and sugarcane have not been benefited to increased carbon dioxide concentration (Mishra *et al.*, 1995).

Increased carbon dioxide concentration increased water use efficiency in the plants (Mishra 1996). In some plants carbon dioxide induces partial closure of stomata through which CO_2 enters the leaf for photosynthesis and through which water vapour simultaneously escapes in transpiration. This partial closure should still allow the same amount of CO_2 to enter, as the carbon dioxide gradient would be greater in a higher CO_2 environment and reduce the water loss of plants. The net effect is the, increase of resistant capacity under water stress condition (and more tolerant of atmospheric pollution as partial closed stomates impede entry of potentially harmful air pollutants into leaves).

Kimball (1982) reviewed 437 experiments published in this last century which dealt with yields of agricultural crops exposed to varying CO_2 concentrations and concluded that doubling of atmospheric CO_2 could increase global productivity by 33 per cent without additional inputs of water and fertilizers. However, weather changes in climate will offset the crop yield grains attributable to direct enhancement of photosynthesis by higher level of CO_2 will continue to be debated (Waggoner 1984).

Sulphur Dioxide (SO_2)

Sulphur dioxide is a colourless gas emitted from combustion of coal and oil. Other major sources are burning of fossil fuels in thermal power plants. Smelting industries and other processes industries as manufacture of sulphuric acid and fertilizers. It can react catalytically or photochemically with other pollutants to form sulphur trioxide (rapidly hydrating to sulphuric acid), sulphuric acid and sulphates.

Table 11.7: Concentrations of SO_2 Producing Injury

Category	SO_2 Concentration (ppm)	
	Short-term Exposure for One Hour Daily	Long-term Exposure for One Hour Daily
Plants	0.5-2.0	0.01-0.05
Animals	0.5-3.0	0.02-0.10
Man	1.0-4.0	0.05-0.20

Sulphur dioxide is often regarded as the traditional pollutant of urban areas. The highest levels of this pollutant occurred during the sulphurous smogs in the industrial cities. The term 'smog' refers to a synthesis of smoke and fog. The fog droplets readily dissolve in sulphur dioxide to produce sulphuric acid. Thereby adding to the potentially harmful nature of the smog. A polluted fog is less readily evaporated by solar radiation than a clean fog, so the duration of the smog or pollution episode may prolong. This smog is socially unacceptable, economically costly, and unhealthy and even dangerous high concentrations (Brodine 1971and Elson 1979).

This gas causes damage to higher plants, forming necrotic areas on leaf. Plants are relatively more sensitive to SO_2 than animals and men. Leaf area of plants collapses under intense exposure of SO_2. There is bleaching to leaf pigments due to conversion of chlorophyll-a to phaeophytin-a. High concentration of SO_2 in air reduced the pH of leaf tissue of some trees increasing the total sulphur contents of leaves and tree bark. In case of wheat exposure to 0.8 ppm of SO_2 with cool smoke for 2 hours daily for 60 days resulted in the reduction of root and shoot length, number of leaves per plant, biomass and total productivity, number of grains per spike and in yield also. SO_2 also reduces leaf area and biomass production of *Dolichas lablab, Cicer arietinum, lens culinaris, Phaseolus aureus* and *Vigna sinensis*. Sulphur dioxide may injure plants in concentration as low as 1-5 ppm. Since sulphur dioxide is absorbed through the leaf stomata conditions that favour on inhibit the opening of stomata. Similarly, affect on amount of sulphur dioxide is observed. After absorption by the leaf, sulphur dioxide reacts with water and forms phytotoxic sulphite ions. They are slowly oxidized in the cell to produce harmless sulphate ions.

SO_2 affects stomatal pores, stomatal frequency and trichones as well as chloroplast structure. The gas is absorbed after passing through stomata and oxidized to H_2SO_4 or sulphate ions. Sulphure dioxide affects mainly the spongy parenchyma. In the middle of the leaf mainly in the areas between the nerves, dry, transparent light-coloured spots appear for example clover, barley, cotton, wheat and apple.

Nitrogen Oxide

In the clear environment it is present in measurable amount of nitrous oxide, nitric oxide and nitrogen dioxide. Oxide of nitrogen is produced by natural processes, including bacterial action in the soil, lightning and volcanic eruptions and by human activities during combustion processes at temperatures higher than about 1000 °C. Nitric oxides and nitrogen dioxide are the most important oxides of nitrogen for pollution studies. It is produced by combustion of O_2 and N_2 during lightning discharges and by bacterial oxidation of NH_3 in soil. Without contact with air and O_2 or even more readily with O_3 to form the more poisonous nitrogen dioxide. NO_2 react with water vapour in the air to form HNO_3 and it combine with NH_3 to form ammonium nitrate (Lee, 1980).

Nitrous Oxide (N_2O)

N_2O levels in the atmosphere are about 0.5 ppm, whereas global level is estimated to be nearly 0.25 ppm.

Nitric Oxide (NO)

The main source of this gas are industries manufacturing NHO_3, other chemicals and automobile exhausts. A large amount of this gas is easily converted into NO_2 by several chemical reactions. NO is also responsible for several photochemical reactions in the atmosphere for the formation of secondary pollutants like PAN, O_3, carbonyl compounds etc.

Nitrogen Oxide (NO$_2$)

This is deep reddish brown gas, which is the only widely prevalent coloured pollutant gas. This gas is the chief constituent of photochemical smog in the industrial cities. NO$_2$ causes health hazards of man, animals and plants. NO$_2$ is highly injurious to plants. There growth is suppressed when exposed to 0.3 to 0.5 ppm for 10-20 days. Sensitive plants show visible leaf injury when exposed to 4 to 8 ppm for 1-4 hours. Low concentration of NO$_2$ suppresses the growth of plants and high concentration causes the bleaching and bronzing of plants like beans, tomato etc.

Ozone (O$_3$)

Ozone layer in the stratosphere protects the life from harmful UV radiations from Sun. Depletion of this layer by the human activities may have serious effects on plant health. Ozone produced in the atmosphere are secondary products of photochemical reactions of nitrogen dioxide and oxygen in the presence of sunlight. Automobile exhaust is also the most important source of ozone.

Increase in O$_3$ concentration near the earth's surface reduces crop yields significantly. O$_3$ enter plants through stomata, it produces visible damage to leaves and decrease yield and quality of plant product. Ozone is injurious to the leaves of plants, when it is exposed for even a few hours at concentrations of 0.1 to 0.5 ppm. When it enter through stomata cause primarily not only palisade but also other cells by disrupting the cell membrane. Affected cells near stomata collapse and die. While necrotic flecks appear, first on the upper side and later on leaf surface. Many crop plants such as alfalfa, bean, citrus, grape, potato, tomato, tobacco and wheat many ornamental and tree plants like, lilac, several pines, and poplar are quite sensitive to ozone. While some other crops are not sensitive in ozone *viz.*, cabbage, peas, peanuts, pepper, beets, cotton, lettuce, straw berry, apricot they are tolerant to ozone, but O$_3$ may dispose plants to insects. At 0.02 ppm it damages the tobacco, tomato, bean, pine etc. In pine seedling, it causes tip burn. In California, USA fruit and vegetable yields have reduced due to O$_3$ pollution.

Other symptoms appears due to O$_3$ are stippling, mottling and chlorosis of leaves, primarily on upper leaf surface spots are small to large, bleached white to tan, brown or black. Premature defoliation and stunting occurs in plants.

Chlorofluorocarbons (CFCs)

Chlorofluorocarbons (CFCs) especially CFCl$_3$ (Freon-11) used mainly as a propellant in aerosol sprays and CF$_2$Cl$_2$ (Freon-12) used extensively as a cooling agent in refrigerators and air conditioners. CFCs are slowly removed from the atmosphere and with residence times in the range of 65-90 years (Brice *et al.*, 1982). Every produce of CFCs are present into the stratosphere due to its slow diffusion rates. Even today closed the emissions of CFCs in the atmosphere. The concentration of CFCs will continue to increase for several years. CFCs have emerged as the compounds which are caused by the depletion of stratosphere ozone. Natural perturbing effects such as lightning, volcanic eruptions, cosmic rays, solar photon events, global circulation changes and variability of solar radiation, all these affect ozone levels. Presence of CFCs, which have strong absorption bands in parts of the infra-red spectrum, may induce their own greenhouse effect (Ramanathan, 1975).

CFCs were invested in 1928 by scientists at general motors who were searching for a less toxic refrigerant than ammonia. CFCs were regarded as wonderful compounds. They are non-toxic, non-flamable, chemically inert, cheaply produced and useful in a variety of applications.

The discovery of stratospheric ozone losses has brougt about a remarkable quick international response. At a 1989 conference in Helsinki, eighty one nations agreed to phase out CFC production by the end of the 20th century but the goal has not been achieved due to non availability of alternative. Known alternatives that exist is hydrochloroflurocarbons (HCFCs) which release much less chlorine per molecules. We hope to develop halogen-free molecules that work just as well and are no more expensive then CFCs.

Fluorocarbons

Fluorocarbons in the atmosphere came from industrial processes of phosphate fertilizers, ceramics, aluminium, fluorinated hydrocarbons (refrigerants, aerosol propellants etc.), fluorinanted plastic, uranium and other metals. The pollutant is in gaseous or particulate state. In minute amounts of fluorocarbons are beneficial helping prevention of tooth decay in man, however, higher levels become toxic to man and plants. On an average, fluoride, level in air is 0.05 mg/m^3. Fluoride enters into plant leaves through stomata. In plants, it causes tip burn due to accumulation in leaves of conifers. The other symptoms of fluorine damage are well defined marginal necroses. It should be mentioned that since plants do not transfer fluorine absorbed from air to the soils, they are good indicators of fluorine pollution. They are not capable of taking up more than 5 mg fluorine/100 g dry weight from the soil, but Gladiolus, peach, plum and pines are the least which are resistant to fluorine. Whereas many kinds of plant susceptible including corn, peach and tulib actively growing especially wet leaves are most sensitive. 0.1 to 0.2 ppb is toxic to plants. Leaf margins of dicots and leaf tips of monocots turn to dark brown, die and may fall from the leaf. Some plant tolerate fluorocarbon upto 200 ppm.

Hydrocarbons

Chief hydrocarbon air pollutants are benzene, benzpyrene and methane. Their sources are motor vehicles being emitted by evaporation of gasoline with carburetors, crankease etc.

Secondary Air Pollutants

Secondary air pollutants are interlinked with NO_2, hydrocarbons and ozone in the atmosphere. In the process of photochemicals, reaction hydrocarbons and nitrogen oxide under the UV-component of light may react with each other and produce toxic air pollutants such as olefine, aldehydes, ozone, PAN, PB_2N and photochemical smog (Grennfelt and Schjodager 1984).

Olefine (Ethyllene)

Automobile exhausts, burning of gas, fuel oil and coal produced directly in the atmosphere. Very low concentration of olefins even few ppb affected the plants seriously. They may cause dropping of petals, plants leaves develop abnormally and senesce prematurely, plants produce fewer blossoms and fruit. At high levels, they retard growth of tomatos.

Peroxyacetyl Nitrate (PAN)

Chief source of atmosphere are automobile exhausts are other internal combustion engines (Gasoline vapors and incompletely burned gasoline + ozone or nitrogen oxide). PAN is a potent eye irritant at about 1 ppm or less. It may persist for more than 24 hours in photochemical smog. Peroxyacetyl nitrate are toxic to plants and cause death of forest trees. Injury has been observed primarily around metropolitan area where large amounts of hydrocarbons are released into the air from automobile. PAN affects many different kinds of plants over large geographical areas. Surrounding the locus of PAN formation due to diffusion or to dispersal of the pollutant by light air currents and PAN blocks hill reaction in plants. It is taken into leaves through stomata and causes injury at concentrations as low as 0.01 to 0.02 ppm. In large urban areas, concentrations of 0.02 to 0.03 are common. PAN attacks preferentially the spongy parenchyma cells which collapse and are replaced by air pockets and give the leaf a glazed or silvery appearance (Jacobson and Hill, 1970). The broad-leaved plants appeared symptoms on the lower leaf surface while small leaf plants shows symptoms on both sides. Young leaf and tissues are more sensitive to PAN. Periodic exposures of leaf to pan cause banding and in some plants cause pinching of leaf margin due to discolouration and death of the most sensitive affected cells. Plants most affected of PAN are Spinach, Beets, Celery, Tobacco, Pepper, Lettuce, Alfalfa, Aster, and Primose.

Photochemical Smog

It is oxidizing pollutant of atmosphere. The word Smog is coined by combining smoke and fog. It prominently occurs where sulphur-rich coal was used. Photochemical smog is produced by the results of photochemical reaction among NO_x, hydrocarbon and oxygen. Photochemical smog formation generally occurred during night or cloudy days. It reduced the visibility. Photochemical smogs are alarming as the chief air pollutant in the industrial cities. 1987 Mumbai suffered with heavy smog for about 10 days. Some sulphates and nitrates also formed photochemical smog due to oxidation of sulphur and damage to plants. Combination of certain environmental pollutant caused synergistic effects in which the injury caused by exposure to factor individually. For instance, when white pine seedlings are exposed to sub threshold concentration of ozone and sulphur dioxide individually, no visible injury occurs. If some concentration of pollutants are given together, however, visible damage occurs (Krahl-Urban, 1988). In case of alfalfa, SO_2 and O_3 together cause less damage than either do alone. These complex interactions are unpredictable for future effects of pollution.

Particulate Matter (PM)

An aerosol is a system of solid particles or liquid droplets suspended in a gaseous medium in the atmosphere and of microscopic or submicroscopic dimensions. Air borne matter results not only from direct emission of particles but also transformations to form particle. This includes dust, ash, soot, lint, smoke, pollen, algal cell, spores and many other suspend materials. Anthropogenic particulate emissions amount to about 100 million metric tones per year worldwide. Atmospheric particulate matter range in size from 0.001 μm to several hundred μm. PM arises from natural as well as man made sources. Natural sources are soil and rock debris (dust), volcanic emission, sea spray, forest fires and reaction between natural gas emissions. Their emission rates are as fallows:

Table 11.8: Estimated emissions of Particulate matter

Sources	Estimated Emissions Tg/yr
Soil and rock debris	50-350
Forest fires	10-50
Sea salt	300
Volcanic emission	25-150
Gaseous emission	345-1100

Particulates often are the most apparent air pollutant since they reduce visibility and leave dirty deposits on the plant leaf surface. They can be directly toxic, damaging sensitive cell membranes much as irritants do in human lungs. With in a few days of exposure to toxic levels of pm mottling (discoloration) occurs in leaves due to chlorosis and then necrotic spots develops, grow poorly and may die. Some dusts are toxic and burn leaf tissues directly or after dissolving in dew or rain water. Biological particulate matter suspended in atmosphere *i.e.* bacterial cells, spores of fungi, pollen grains, causes many diseases plants.

Toxicants

There are more than 70,000 toxic chemicals available today which are air pollutants, causes human and plant health hazard. Although the acute effects of many of those chemicals have been recognized, but others have no knowledge of their potential chronic effects on man, animal and plants. Some of principal toxicants are as follows –

☆ **Arsenic-** is produced as a by-product of metal refining process.

☆ **Dioxin-** Trichlorophenol made, as herbicide released chemical dioxin. These chemicals spread in the air in 1979, within two weeks animals and plants were dying and people were admitted into the hospital for skin lesions and vomiting.

☆ **Asbestos-** Asbestos is a mineral fibre used in cement pipes, flooring products, paper, asbestos roofing products, which are non-degradable. They cause many alterations to the plants.

☆ **Carbon tetrachloride and chloroform-** Carbon tetrachloride and chloroform use for making fluorocarbons for refrigerants and propellants etc.

☆ **Isocyanate** is used in the manufacturing of plastics, dyes and pesticide.

☆ **Polychlorinated** biphenyls are used in epoxy paints, protective coating for wood, metal and concrete. It is also used in coolants and insulators in high voltage transformers.

☆ **Chromium** is used in stainless tool and alloy steel, heat and corrosion resistant materials.

☆ **Vinyl chloride** is used to make polyvinyl chloride (PVC), a plastic which is used for many purposes including food, wrappings, water piping.

☆ **Nitrosamines** are mostly used in rubber processing organic chemical manufacturing and rocket fuel.

☆ **Polycyclic organic hydrocarbon (POH) and Polycyclic aromatic hydrocarbon (PAH)-** These organic materials are known to carcinogenic effects in animal and man.

Many other toxicants are also released in the atmosphere need to be undertaken to identify chemical which may cause chronic adverse effects before these pollutants become widely distributed in the atmosphere.

Toxic Metals

Many toxic metals are mined and used in manufacturing processes or occur as trace elements in fuels, especially coal. It is also released from industries and human activities in the atmosphere. The common metals are present in the air are mercury, lead, zinc cadmium, nickel, beryllium, thallium, cesium and plutonium (UK Royal Commission on environmental pollution, 1983).

☆ **Lead:** These metals are released to the air in the form of metal fumes or suspended particulates by fuel combustion, or smelting and disposal wastes. Worldwide lead emissions amount to about 2 million metric tones per year or two-thirds of all metallic air pollution. Most of this lead is from leaded gasoline. Lead is metabolic poison that binds to essential enzymes and cellular components and inactivates them.

☆ **Mercury:** Mercury is another dangerous and wide spread environmental pollutant. It is a liquid volatile metal found in rocks and soil is present in the air as results of human activities as the use of mercury compound, production of fungicides, paints, cosmetics, paper pulp and cool burning power plants. Long-range transport of mercury through the air is causing bioaccumulation in aquatic ecosystems far from the source of emission.

☆ **Zinc:** It is found in the air around zinc smelters and scrap zinc refineries, copper, lead and steel refineries release some zinc in the air. Zinc in air occurs mostly as while zinc oxides fumes and is toxic to man and animal.

☆ **Cadmium:** Cadmium is emitted as vapour in the atmosphere. In this state it is quickly reacts and form oxides, sulphate or chloride compounds. Cadmium is poisonous at very low levels and is known to accumulate in plant and animals, it is cause serious diseases. Major sources of cadmium are industrial extraction, refineries, electroplanting and welding of cadmium containing materials. Production of pesticides and phosphatic fertilizers also produces cadmium in the air.

12

Environmental Pollution Management through Microbes

Rapid increase of human population will create food scarcity in the future. This is a great concern to scientists and farmers for utilization of chemicals for more production. The chemical compounds are known to cause pollution. The chemical compound *viz.*, fertilizers, fungicides, insecticide, nematicide etc. are also polluting the soil and ground water, in addition to chemical use in the agriculture. Other potential dangers of environmental pollution by heavy metal compounds metaloids, radionuclides organomental and related substances due to industrialization and urbanization have generated new problems. During the last 50 years of industrial development, a number of hazardous substances have been added to the environment (soil and water). There are an estimated 1,00,000 human-made chemical in use and hundreds of new ones are produced each years (Gupta and Mukerji, 2003 and Sandbach, 1982). Increased urbanization and industrialization have led to generation of enormous amounts of effluents containing heavy metals several folds in excess than the permissible limits. Unlike other pollutants the metals due to their recalcitrant nature tend to accumulate in various forms of life and enter the food chain causing environmental hazards (Silver, 1981). It is well perceived that there is a permissible limit of each metal and above which they are generally toxic (Gadd, 1992 a). Soil fertility and the earth's recuperative power areas damaged in the end by the excessive use of chemical fertilizers and by deep mechanical ploughing which damage the top soil through over exposer. Pesticides and chemical fertilizers are also big pollutants these tends to pollute the soil and agricultural produces and rendering them unfit for human and animal consumption. Their persistent use in

Japan is reported to have poisoned land and water surrounding major cities. In this scenario of increasing demand for food and their diminishing production, the widespread hunger famine or starvation endangering global peace especially in Africa and Asia.

The non-renewable mineral resources of the world are also fast being depleted. According to study undertaken by the United States' National Academy of Sciences, the world mineral resources like iron and tungsten, may not last beyond two hundred year at the present rate of utilization. The greatest pollutant today is radioactive fall out from nuclear testing or form long-term storage and disposal of nuclear wastes. When, in 1954, a hydrogen bomb was tested by the United States in the pacific area, it caused widespread radiation, sickness among a large number of Japanese fisherman and as test continue (with the Chinese and French still conducting in the atmosphere) which also caused widespread radioactive pollution.

It is widely and quite reasonably believed that the cumulative pollution of the atmosphere, land and water may in the long run interfere with the life support systems. Thus, it has become necessary for man to direct his intellectual power not only at technological development but also towards the protection of natural life support system by biological treatment of contaminated soils and waters. In an undisturbed ecosystem, insects and weeds are to a great extent controlled by their natural enemies without disturbing ecosystem. Pest can be controlled by releasing more predators and parasitic insect's mites' fungi etc. Biological control involves:

1. Conservation of natural enemies
2. Release of parasites
3. Use of microbial agents
4. Use of predators

Conservation of Natural Enemies

The conservation and enhancement of natural enemies should be the first consideration. Conservation means the avoidance of measures that destroy the natural enemies and enhancement in use of measures that increase their longevity and reproduction attractiveness of an area to natural enemies. Preservation of natural enemies on inactive stage is also a problem when the reservoir is small. Pupae of Epipyrops are found in large numbers on the thrashes of sugarcane leaves at the time of harvesting if the leaves are not burned and left in the field. The adults emerge to augment the supply of natural enemies in the premonsoon season against *Purill perpurilla* on the young crop of sugarcane. The concept"if more the diversity more is the stability" hold true in conserving and enhancing the natural enemies.

Release of Parasites

Natural enemies control the agricultural pests where the damage was above the threshold levels. In this situations, the numbers of natural enemies usually have to be raised artificially by mass production in the laboratory. This means that scientists have to develop an artificial diet and work out techniques of mass rearing. This method of control is cheap and environmentally safe.

Use of Microbial Agent

The number of microbial agents such as bacteria, fungi, viruses, Protozoa and nematodes have been found infecting insects. Microbial agents should be harmless to other forms of life, easy to produce, store and apply through conventional sprayers and dusters. Among the bacteria, *Bacillus thuringiensis* was found promising against a number of lapidopterous larvae. Strains were developed which produced maximum amount of toxicant and is now considered reliable microbial control agent.

Use of Predators

Insect predators are commonly found the *Coccinellid* predators, *Chiloconus, Pharoscy mnus, Cryptolamus, Scymnus* and *Monochilus* feed on melay bugs *Coccids, Scales* and mites on citrus grapevine and guava. The miride predators *Crytorhinchus* attack brown plant-hopper and green leafhopper of rice. The green lacewing *Chrysopa* another most effective predator feeds a melay bugs and aphids. The young caterpillar spiders also help in insect control. The walf spider, *Lycosa* is important in the rice ecosystem.

Table 12.1: List of Antagonistic Microorganisms Use of Biocontrol

Sl.No.	Pathogen	Name of Disease	Antagonistic Microorganism
1.	*Heterobasidion annosum*	Root or bud rot of conifers	*Peniophora gigantea*
2.	*Endothia parasitica*	Chestnut blight	s-RNA of same fungus
3.	*Botrytis cinerea*	Post bloom dead of tomato	*Gladosporium herbarum Penicillium* sp.
4.	*Nectria galligena*	Canker	*Trichoderma* sp.
5.	*Cystospora* and *Botrytis*	Botrytis rot of straw berries	*Trichoderma* sp.
6.	*Sphaerotheca fulginea*	Cucumber powdery mildew	*Ampelomyces quisqualis*
7.	*Puccinea recondita*	Wheat leaf rust	*Darluca filum*
8.	*Penicillium digitatum*	Citrus green mold	*Trichoderma viride*
9.	*Phytophthora cinnamorri*	Pine seedling	*Mycorrihizal fungi*
10.	*Fusarium oxysporum*	Tomato seedling	*Mycorrihizal fungi*
11.	*Verticillium* sp.	Cotton wilt	*Mycorrihizal fungi*
12.	*Agrobacterium tumefaciens*	Crown gall Pome, grapes stone	*Agrobacterium radiobacter* strain K-84
13.	Seed Treatment	Seed born disease	*Bacillus subtilis* strain A-13 *Streptomyces* sp.
14.	*Monilinia fructicola*	Brown rot	*Pseudomonas putida*
15.	*Erwinia amylovora*	Fire blisht of Apple	*Erwinia herbicola*
16.	*Xanthormonos translucens*	Bacterial leaf streak of rice	*Erwinia* and *Pseudomonas*
17.	*Cercospora* sp.	Leaf spots	*Pseudomonas cepacia*
18.	*Alternaria*	Leaf spots	*Bacillus* sp.

Reduced use of total 90,000 tonnes of annual pesticide (Fungicide and insecticide) for agricultural purpose and nearly 600 crore rupees agro-pesticides marketed is concern. The Biological control of plant disease is practised for saving money and environmental pollution. *Streptomyces scabies* caused disease of scab in potato has been control by the inoculation of *Streptomyces praecox* in soil (Millard and Taylor, 1927). *Peniophora gaianteae* is also effective competition of *Fomes annosus* caused root disease of pines (Rishbeth, 1963). Thus, biological control in any condition is practised and reduced the incidence of the disease caused by the pathogens (Table 12.1).

Biofertilizers in Agriculture Productivity

Chemical fertilizer constitutes a serious problem and farming does not remain economically feasible. Moreover increased used of fertilizer lead to health hazards and soil erosion. Nevertheless, need of more food production for increasing populations are taking necessary action for use of biofertilizer.

The most important of the biological processes existing on earth are photosynthesis and nitrogen fixation. Biological nitrogen fixation is second to photosynthesis nitrogen as one of the most abundant elements in the biosphere helps in soil fertility. In the process of nitrogen fixation, the atmospheric nitrogen is converted into ammonia, which in most instances is rapidly assimilated by the metabolic process of the cell into organic nitrogen containing components. The estimated fixation between 100 and 200 million tones of nitrogen every year represent more than 70 per cent of the input into the soil and water nitrogen.

The food requirement of increasing population, the nitrogen fertilizer requirement is about 11.25×10^6 tonnes as against 3.9×10^6 tonnes at present. The vast gap of nitrogen fertilizers can be fulfilled by the use of biofertilizers, which is sustaining agricultural productivity. Application of biofertilizers in the fields is the viable alternative. Biofertilizers is a living fertilizer composed of microbial inoculants or group of microorganisms, which are able to fix atmospheric nitrogen (Panwar and Sirohi, 1989). Free living bacteria (*Azotobacter and Azospirillum*) and the blue-green algae and symbionts such as *Rhizobium* (Carling *et al.,* 1978). Association of Azolla and blue-green algae *Anabaena* is also well known.

Rhizobium Biofertilizer

For environmentally sustainable agriculture development, the *Rhizobium* is the best biofertilizer, which fixes atmospheric nitrogen symbiotically with legumes (Mishra, 1992). All the *Rhizobium* is abundant in the soil but not all are able to nodulate in all type of legumes due to their specificity. Although, effective inoculum for each legumes is available. The country's requirement is about 11.25×10^6 kg per year. Inoculant for more than two dozen legume crops have been developed. In India approximately 30 million hectares of land is under cultivation of pulses and legumes. Effective inoculants have been found to supply upto 60 kg N hectare. In general 20-40 per cent increase in the crop yield is recorded (Mishra, 1992). Small or marginal farmers have little or no access to chemical fertilizers. This appears to be wide scope for the development of biofertilizer industry in India and abroad.

Azotobacter, VAM and Azospirillum Biofertilizer

Azospirillum brasilense and VAM (*vesicular Arbuscular mycorrhiza*) fungi showed beneficial effect on plant growth and yield (Okon, 1985). VAM fungal hyphae have extra absorbing surface and reach beyond the soil zone explored by the root hairs (Bogyaraj, 1990). Inoculation with VAM fungi, single or in combination with *Azospirillum* brought about high increase in chlorophyll content and maintained it. The rate of photosynthesis was higher in plants inoculated with either VAM or *Azospirillum*. The synergistic effect was also noticed with dual inoculation (Panwar *et al.*, 1990). Vascular-arbuscular micorrhizal symbiosis can affect stomatal behaviour of host leaves (Bethlen Falvay *et al.*, 1988, Panwar, 1991) because the mycorrhizal root system are capable of more water uptake than non-micorrhizal root systems. The *Azospirillum*-plant root association has also been reported to improve crop productivity by altering root-shoot ratio (Okon, 1985).

Blue-green Algae Biofertilizers (BGA)

Blue-green algae are capable of nitrogen fixation. Most of the species have nitrogen fixing ability to the order *Nostocales* and stigonematales more than 100 species of BGA are known to fix atmospheric nitrogen. These have been found to be very effective in rice and banana crops. The BGA increased the grain yield of rice about 586 kg/hectare. India is one of the countries where agro-ecological conditions are favourable for using blue green algae technology in place of chemical fertilizers. In some parts of our country, production of BGA inoculants has commercialized.

Pushpa Srivastava (1999) was work done on effluent on textile factory and reported tolerant microflora as a biological source of their treatment. The emphasis has been on algal flora, they are often useful means of water purification. Sewage and water are treatment rendering the reclaimed water suitable for agricultural and industrial purposes. Cyanophyta (blue-green), Euglenophyta, chlorophyta (greens) and Bacileariophyta (Diatoms) assist in the purification of atmosphere by consuming organic nutrients and carbon dioxide released by other concomitant non-photosynthetic microflora and produce oxygen rendering water fit for irrigation, which purify the habitat though their photosynthetic activity. Therefore, it is proposed, that proper tanks be constructed, where effluent should be collected. It should be treated before it is recycled for agricultural, washing and bathing purposes. Biological treatment of the effluent with the help of tolerant algae could be the cheapest and the best method (Rana, 1975, Raj, 1992 and Witton, 1994) for recycling of the effluent. This method has been adopted and operated successfully by Saras Dairy for its effluent treatment (Sharma and Srivastava, 1994). The effluent was collected in large pond. It was pushed through a fountain, thereby the effluent came in contact with atmospheric oxygen raising the biological activity of the microorganisms present in effluent. This led to the enormous growth of the effluent tolerant algae. After sufficient growth of the algae was skimmed off manually and the water was filtered and stored into another pond. The water is proposed to be used for developing garden and algae as biofertilizer.

Eight species of heterocystous Cyalno bacteria belonging to the genas *Anabaena*, *Nostoc*; *Calothrix* and *Aulogira* were isolated from the rice field of Orissa state, which have nitrogen fixing ability, where the pesticide application were done (Sahu *et al.*, 1986). In order to find out the pesticide tolerance capacity and increase their growth response and photosynthetic oxygen evolution in different concentration of Balwan and Thiokill pesticide.

Azolla Biofertilizer

Azolla has gained worldwide recognition as biofertilizer particularly due to the concentration of *Anabaena azolla* that fixs molecular of N by the process of symbiosis. *Azolla* is a floating aquatic fern and its nutrition, nitrogen fixation and growth depend on different nutrients more directly in floodwater than in soil (Ali and Watanable, 1991). Lumpkind Plucknett (1982), Watanable and Ramirez (1984) and Kushari and Watanable (1991) have conformed the effect of P in the growth of *Azolla*. Mallick and Kushari (1996) found that *Azolla* accumulated maximum K in 40-50 ppm concentration of nutrient solution and he suggested that K have a key role in the growth and nitrogen fixation of *Azolla*. This biofertilizer is used for rice cultivation in different parts of world such as Vietnam, China, Thailand, Phillipines and India. It is also used in fish culture ponds. Azolla can easily grow in cooler regions. Field trial indicated that rice yields are increased by 0.5-2 tonnes/h due to use of Azolla. There is need to develop tolerant strains to high temperature salinity and pest and disease resistance for wide adoptation of Azolla biofertilizers.

Phosphorus Biofertilizers

The root fungus association or mycorrhiza has high potential to accumulate phosphorus from the atmosphere in the plants. Mixture of charcoal and soil is necessary material for the commercial growth of these inoculants. It is reported by many scientist that microphos cultures increase yield upto 200-500 kg/hectare and save 20-40 kg superphosphetic fertilizers. It is environmentally sustainable and economically feasible.

Management of Metal Pollution

Over past century, restricted mining, extensive industrialization, large-scale urbanization, modern agricultural methods and faulty waste disposal practices have resulted in release of unprecedented levels of toxic pollutants including heavy metals in the environment. Guckerts (1987) described the affinity under the metals was Pb, Cu, Cd, Zn for high molecular weight compound exudates from the seveage irrigated water. The effect of different levels of lead on yield and utilization of nutrients by spinach crop was studied by Vinay Singh and Surendra Singh (1984) an sandy loam soil, laed application decreased the growth yields and concentration of NPK, Ca, Mg and Mn in spinach plants at all stages of growth. The heavy metal containing residue of industrial sewage sludge used for irrigation, decrease microbial population *i.e.* *rhizobium* and other beneficial bacteria, it also increased harmful fungal and bacterial population in soil rhizosphere (Mishra, 1992). Therefore, risks arising out of uncontrolled release of heavy metals have become so clear that it is no longer acceptable to allow the metal-laden wastes in soil and water. The conventional techniques of treatment is not only expensive but also generate fresh pollutants.

Microorganisms have contributed significantly to the formation and decomposition of minerals in the earth crust since geologically ancient times. Many bacteria have helped the mining operations as they could leach the metals from insoluble. Although the basis of bioleaching was not known till the date 1000 B.C. The welsh in 16th century and Spanish in 18th century used microbial leaching for recovery of heavy metals. Biological leaching is now being successfully used in many countries to recover metal *viz.*, copper, uranium, cobalt, nickel, zinc, lead, silver and gold. According to an estimate made by Gorham International Inc., the annual production of the worldwide microbial metal industry will amount to 90 billion dollar by the year 2000.

The ability should be in microorganism, which are used in leaching process to grow and survive under metal polluted conditions, is frequently encountered in natural and is attributed to stress. Bacteria *Thiobacillus ferroxidans* was isolated and characterized for leaching of metal sulphide ores (Bellivean *et al.*, 1991). Many other species are *T. thiooxidans, T. acidophilus* have often been used together with *T. ferroxidans* in industrial leaching processes. Other bacteria identified such studies include *Leptospirillum ferroxidans* and *sulpholobus* and other thermophillic archae bacteria such as *Acids anus brierley* and some sulphate reducing bacteria. This would open up immense possibilities for employing manipulated forms of these bacteria in the bioleaching process (Merglay, 1991).

Microorganisms and microbial products are highly productive bioaccumulation of soluble and particulate forms of metals especially from dilute external concentrations. Such noval capabilities of microorganisms are of industrial interest because the biodegradation or removal of potentially hazardous heavy metals and radionuclides from industrial effect and wastewaters can serve dual purpose that will lead to both detoxification and thus environmental restoration and recovery of precious elements including gold and silver (Brietey *et al.*, 1985, Gadd and Griffiths, 1978). The microbe-based processes after low operation cost, minimization of the volume of chemical and/or biological sewage/sludge to be disposed of high efficiency detoxification of very dilute effluents. Microbial related biosorption technologies would provide an environment friendly alternative metal removal and recovery (Wase and Forster, 1997), Ahuja and Tiwari (2003) has been found, a bacterial isolate *Acinetobacter anitrotus* absorb 98 per cent of Pb and 85 per cent of Cd from dilute solutions of heavy metals. This heavy metal biosorption phenomenon was swift *i.e.*, occurring with in a minute of reaction time.

Today application of microbial biomass in this area has occupied a place of prominence due to easy availability and economic production of microbial biomass. The microbial biomass, phototrophs *i.e.*, algae and Cyano-bacteria offer a versatile and flexible system for a wide range of applications. Rajni Gupta and Ahuja (2003) reported the Cyanobacterial group have good metal adsorption capacity and he found *Oscillatoria* to passes excellent capacity to bioabsorb copper, zinc, cobalt at an equilibrium concentration of 120 ppm.

Number of synthetic compounds, which are not related to natural ones persist in the environment and create health hazards for human beings. The contaminated land sites are unsuitable for housing and agriculture. Bioremediations is the productive use of microorganisms to remove or detoxify pollutants usually as contaminants of soil, water etc.

Hydrocarbon oxidizing microorganism constitute from 5-50 per cent of total microbial population in oil-polluted area (Maccubbin and Howard, 1979). Several hundred strains of acquatic hydrocarbon degrading bacteria have now been identified. The most prevalent genera are *Pseudomonas micrococcus, Enterobacter, Vibrio, Flavobacterium, Bacillus* and *Micobacterium*. Fungi like *Candida cladosporium, Torulopsis* and *Penicillium* have shown-hydrocarbon degradation. Complete degradation depends upon complex inter play between many microbial species (Garg *et al.*, 1994). Pesticide (Insecticide, Fungicide and nematicide) are perhaps most extensively used in agricultural practices. Microbial degradation through hydrolysis of P-O alkyl and P-O aryl bonds are considered as the most significant steps in detoxification of organophosphorus compounds. Phosphotoriesterase is the most important enzymes in the bacterial metabolism of organo-phosphates. Recently *Pseudomonas putida* was reported to metabolies by P-nitrophenol to hydroquanone and 1,2,4- benzenetrial even the final ring compound was also cleaved by benzene triol oxygenase to malylacetate (Rani and Lalitha Kuroosi, 1994). Biological treatment of nitrogenous fertilizer waste is carried out to remove the pollutants. It can be done by any conventional methods (Nyampfene and Mtetwa, 1987). The biological treatment of the effluent is achieved in three different parts of (1) Hydrolysis of nitrogen from urea (2) Nitrification of ammonical nitrogen (3) Dinitrification of nitrite and nitrate.

13

Agriculture Waste Management through Composting for Mushroom Cultivation

India is known as an agricultural country with more than 70 per cent population living in villages having agriculture as the main occupation. Proliferation of green revolution in mid sixties, the overall production of food grain has risen and this time total cropped sown area goes to more than 160 million ha. Increasing the food production also increased Agricultural Wastes, which was major problem to utilize it for further production. Mushrooms as such fall in this category, which require much agricultural wastes. In the year of 1961 Government started cultivation of mushroom through the recycling of agricultural waste with the use of modern technology. *Agaricus* species popularly known as white button mushroom is most important commercially among all the cultivated mushroom. The fungus is saprophyte and heterotrophic requiring complex organic compounds preformed by other microorganisms. The carbon compounds are syntherized by microorganisms in Agricultural straw such as Wheat, Paddy, Brassica, Sugarcane, Maize, Bajra, Jawar, Pea, Gram, Oat, Arhar (Pods leaf) tree leaf and grasses. The main function of these materials is to provide a reservoir of cellulose, hemicellulose and lignin, which is utilized by mushroom for its growth (Vijay, 1996). These materials also provide proper physical structure to the substrate mixture to ensure the necessary aeration for build up of the mycoflora isolated by Mishra *et al.* (2007). During the compost preparation the natural mesophilic flora subsides and increase of thermophilic microflora build up in the pile. Microflora of the compost at various stages has been studied by various workers. Hayes (1968) isolated 44 species of mesophilic fungi from phase first compost *Absida cylindriospora, Mucor hiemales, Mucor*

Thamnidium elgans and *Zygorynchus moitler* active upto third day of composting after that *Aspergillus repens, Copulariopsis brevicaulis, Cladosporium, Cladosporioides, Curbularia lunata* were active. Mishra *et al.* (2007) isolated mycoflora of compost as *Scytalidium thermophilum, Humicola isolens, Tolaro myces emersoni, Chaetomium indicum, Mucor racemosus, Aspergillus* sp., *Penicellium Trichoderma, rhizopus, Verticillium, Alternaria, Alelminthosporium, Cladosporium, Trichothecium rosaum, Scopulariopsis* and *Papulosporia byssina*. Population dynamics of different thermophylic fungi determined at 0 day, 6th day, 9th day and 12th day of incubation at 50°C. Rewardand Caillux (1972) also checked the fungal population at different times, levels and temperature of incubation and indicated succession of species. Similarly Stratsma *et al.* (1949a) isolated twenty different thermophilic fungi from the compost and indicated that *S. thermophilum,* which is an important organism for compost preparation.

Agricultural waste materials now converted in the compost and is ready for filling in the bulk chamber, which can be done manually by employing labour or mechanically by using a conveyor or it is sterilized by chemical and ready for growing mushroom.

Most parts of India, the growers do not have the finances to afford the pasteurization units at individual basis. Therefore, there is a need of co-operations which can produce compost for a number of growers, pasteurize it and supply spawned compost and pasteurized casing material to growers at a marginal profit for sustainable environment and converting various wastes into a prized product.

14

Environmental Pollution Management through Biodiversity

Development of any country is directly by related to development of agricultural practices and industrialization of land. This craze resulted into unlimited distraction of natural resources, which had adverse effect on all forms of living organisms in the biosphere. This situation was created by human itself, threatened the survival not only of man but also all kinds of living organism. In the near future, every public place at you will see the warning that "Air unfit for breathing", "Water unfit for drinking", "Do not eat fish caught here" and so on. Thus environmental pollution should be managed at any cost for future generations through the management of Biodiversity.

Biodiversity, as this assemblage of life forms is referred to has now been acknowledged as the foundation for sustainable livelihood and food security. The United Nations Conference on Environment and Development (UNCED) held in Rio de janeiro in June 1992 was "Convention on Biological Diversity" which was signed by 156 countries and European community helped to place the loss of biodiversity and its conservation on the global agenda, resulting the biodiversity becoming a household word. Scientists have estimated that more than 50 million species of plants and animals including invertebrates and microorganisms occur on earth and hardly 2 million of them have been described by man so far.

Biodiversity in the World

In the world, about 30 million insects 15,210 mammals reptiles and amphibians, 9,225 birds, 21000 fishes, about 4,80,000 plants and 3 million other invertebrates and microorganisms are found. Many of them have not been identified. Most of the 1,700 million hectares of tropical forest, rich in biodiversity were located in poor countries, while such forests covered barely 7 per cent of the land surface. They harboured half of the species of the world's flora and fauna. Much of the world's agricultural and pharmaceutical needs from developing hybrid seeds to herbal cures came from such prime forests. Famous Harvard biologist E.O. Wilson estimates that the chopping down of tropical forests leads to extinction of at least 50,000 invertebrate species every year, about 140 face extinction every day.

Biodiversity in India

India has nearly 50,000 plant and 80,000 animal species already identified and described. There are about 20,000 flowering plants and 67,000 insects, 1000 mollusces, 6500 other invertebrates, 1,400 fishes, 140 amphibions, 420 reptiles, 1200 birds and 340 mammales. The country have important vavilovian centre of Diversity and origin of over 167 important cultivated plant species and some domesticated animals.

India is recognized as a country uniquely rich in all aspects of biodiversity ecosystem species and genetic for any one country in the world. It has largest array of environmental stipulations by virtue of its tropical location, varied physical features and climatic types.

The Indian region alone has given to the world nearly 167 economical plants whose centre of origin/diversity lies in India example Rice, Sugarcane, Minor Millets, Brassicas, Rice-bean Asiatic vignas, eggplant, Banana, Citrus, Mango, Cardamom Jackfruit, Jute, Edible diascorea, Black pepper, amaranthus, turmeric, ginger, cucurbits and many herbal drugs.

Important

Broadly, biodiversity satisfied human needs in two different ways, direct and indirect.

Biodiversity will help not only in increasing agricultural productivity but also developing disease resistant varieties, which save man and environment both. It was evident in the early 1970's when an epidemic called grassy stunt virus destroyed more than 1,60000 hectare of rice in Asia, it could be controlled from single sample of wild rice *Oryza nivra* from central India, which was found to be the only resistant genetic resource of grassy stunt.

Besides food and other basic needs of human health has gained priority in welfare programme. Medicinal plant resources are supplied in the world market which gained worth of 40 billion dollars every year.

Indirect benefits include nutrient trapping, maintaining water cycles, soil production, protection of soil absorption and break down of pollution.

Conservation of Biological Diversity

Biological diversity can be conserved by two ways, *in situ* and *ex situ*. *Ex situ* conserved germplasm banks and seed stores *in vitro* collection. It is maintenance of species provided by botanical gardens, zoos and aquarium and grass root collection of plant cultivators and animal breeds. *In situ* conservation is done by protecting areas rich in biodiversity. These include biosphere reserves, national parks and sanctuaries. The primary objective of this conservation is to save genetic variability for future use and maintaine balanced ecosystem.

It is assumed that in the next 20-30 years the world could lose more than a million species of plants and animals. This is cause of environmental change due to human activities. Most of agricultural crops currently being cultivated have been selected for a particular geographic area, its productivity may not be viable if the climatic condition changes and new pests and diseases attack. For this purpose more and more genetic diversity is needed to maintain food security in the new climatic condition.

GLOBAL WARMING

There is a large number of simple and complex models which quantified the greenhouse effects at atmospheric climate. The majority of the model warned for global warming of $3\pm1.5°C$. The US National Research Council (1979) predicted that the lower latitudes will experience an annual mean temperature increase between $1.5\pm3.0°C$ while the polar and subpolar latitudes will experience a warming of $4.0-8.0°C$.

Greenhouse Effect

Carbon dioxide concentration of pre-industrial world was 300 ppm and a future world will experience a carbon dioxide concentration of 600 ppm by the year of 2075. Carbon dioxide was released into atmosphere at a faster rate than the capacity of oceans to absorb it. Thus its increased concentration, the thick layer of this gas prevents, the heat from being reradiated out. This thick CO_2 layer function like the glass panels of a greenhouse (The glass window of Motorcar), allowing the sunlight to filter through but preventing the heat from being reradiated in outer space. This is called greenhouse effect. Most of the heat is absorbed by CO_2 layer and water vapoures in the atmosphere which adds to the heat that is already present in the atmosphere. The net result is the heating up of the earths atmosphere.

CO_2 increases the earth temperature by 48 per cent while CFCs are responsible for another 21 per cent increase. Other gases which contributed greenhouse effects are SO_2, NO_x, CH_4, NH_3 and H_2S-discharged by industry and Agriculture (Figures 15.1 and 15.2).

The earth's mean temperature will be 1.5 to 4.5°C by 2050. According to model projection changes will be the least in the tropics and the most at the poles.

Hanson *et al.* (1983) believe that the CO_2 greenhouse warming should emerge from the noise level of natural climate variability by the end of the century and there is a high probability of warming in the 1980. Indeed, the northern Hemisphere mean temperature decrease which began in the 1940 has been arrested and even reversed during the mid 1970. However, other climatic influences such as volcanic eruptions may yet counteract and obscure a CO_2 induced warming trend such that undisputed confirmation of its detection may not be available for some time.

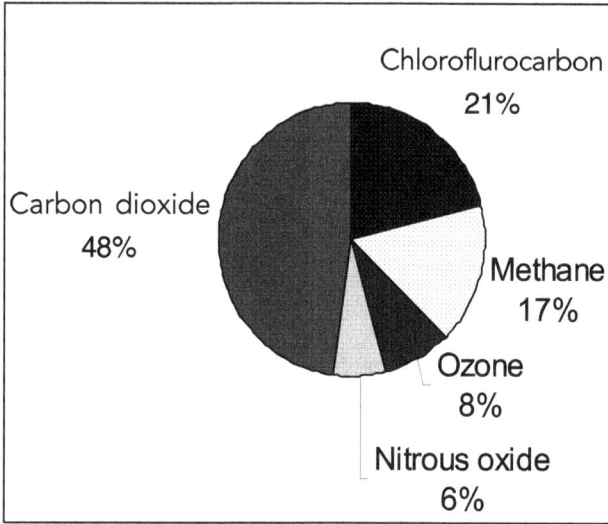

Figure 15.1: Contribution of Global Warming
Source: World Resource Institute and UNEP.

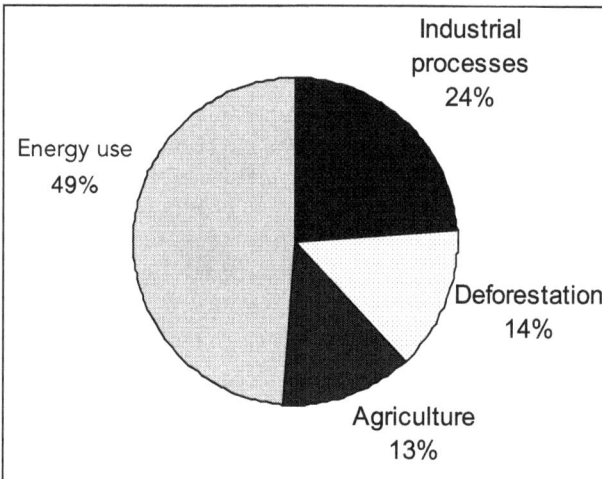

Figure 15.2: Contribution to Global Warming by Human Action.
Source: Data from World Resources Institute.

A rise of five degree surface air temperature would raise the sea level by five meters within few decades and threatening the all densely populated coastal cities from Shanghai to San Francisco. It is also suggested that northern America would be warmer and drier. Therefore the U.S. produce less grains. On the other hand, North and East Africa and middle east India, West Australia would be warmer and wetter and produce more grain by the increasing of rice growing areas as well as season also.

Bach (1978) studied data on weather and corn yield in the united states corn belt states from 1901-72 and estimated that corn production in those states would change by about 11 per cent for every 1°C change in average maximum temperature. Own the growing season, and by about 15 per cent for each 10 per cent change in precipitation. He concluded that warmer and drier weather would decrease corn production in the corn belt and cooler and wetter weather would increase it.

A rise in sea level of 50-100 cm caused by ocean warming would flood low-lying lands in Bangladesh and West Bengal due to greenhouse effect. This may lead to more hurricanes and cyclones and early snow melts in mountains causing more floods during monsoon. If unchecked release of greenhouse gases remains, it could alter temperature, rainfall and sea level of the earth. Therefore, UNEP has appropriately chosen the slogan "Global Warming". G-8 country agreed to reduce 25 per cent release of greenhouse gases. This agreement was done in the meeting of G-8 countries held at Tokyo, Japan in 2008.

Growth of phytoplankton (Single celled photosynthetic organisms) in the ocean might absorb most of the additional CO_2 and balance world temperatures. This could increase productivity of marine ecosystems and make more sea food available for human consumption. There seems to be evidence of increased CO_2 uptake. Between 1991 and 1994 CO_2 increases in the atmosphere were about 7 billion ton per year less than emissions. Some scientists believe that the "Missing CO_2 is being taken up by increased photosynthesis in boreal forests."

OZON DEPLETION

Ozone is a pale blue gas, mostly present in the ozone layer or ozonosphere (10 to 50 km above the earth surface). The ozone layer has two important and interrelated effects. First, it absorbs Sun's ultraviolet radiation, thus it acts as a protective umbrella to the living things on the earth. Second by absorbing the UV radiation, the ozone layer heats the stratosphere causing temperature inversion, it limits the vertical mixing of pollutants thereby causing the dispersal of pollutants over larger areas and near the earth's surface. That is why a dense cloud of pollutants usually hangs over the atmosphere in highly industrialized areas causing several unpleasant effects. The wastes spread horizontally fast reaching all longitudes of the world in about a week and all latitudes within months. Therefore ozone problem is global in scope.

In 1985, the British Antarctic Atmospheric Survey announced a starting and disturbing discovery. Ozone levels in the stratosphere over the South pole were dropping during September to October every year as the sun reappears at the end of the long polar winter. This ozone layer is depleting since 1960 but was not recognized. In 1993 as much as 70 per cent of Antarctic Stratospheric ozone was destroyed over an area about the size of North America. Expert of the Environmental Protection Fund (EPF) believe that within the mid century the excess of solar ultraviolet radiation can destroy 3 per cent of world total production of wheat legumes etc. It was calculated that 1 per cent reduction leads to 6 per cent increase in skin cancer causes a must take care of the ozone. Cold temperatures (-85 to -90°C) in Antarctica play a role in ozone losses. During the long, dark winter months, the strong circumpolar vertex isolates.

Antarctic air allows stratospheric temperatures to drop low enough to create ice crystal at high altitudes, something that rarely happens elsewhere over the world. Ozone and chlorine containing molecules are absorbed on the surfaces of these ice particles. When sun returns in the spring and provides energy to liberate chlorine of ions. One suspected of being most important in ozone losses are chlorofluorocarbons (CFCs) and halogen gases. CFCs enter the stratosphere and remain there for year until they are converted to other products or are transported back to the stratosphere. The stratosphere could be regarded as sink. But these pollutants react with the ozone and deplete it. Use of CFCs was estimated that US release highest, the 37 per cent of the chloroflurocarbons in the atmosphere followed by EEC countries 35 per cent and 11 per cent by Japan. Only 5 per cent is attributed to the developing countries of Latin America, Asia and African country.

In 1975, it was believed that a large fleet of supersonic air craft flying at altitude upto 25 km would have a noticeable effect on the ozone layer (World Health organization 1976b). By 1982 stratospheric flying air craft whether supersonic or subsonic would be expected to have a negative effect on ozone that is causing a significant depletion of the ozone layer (Hulm, 1982). McElroy *et al.* (1976) suggested that reductions in ozone of 25 per cent might result from use o f chemical fertilizers. In contrast Lie *et al.* (1976) estimated depletions of ozone layer is only 1- 4 per cent.

Reinsel (1981) studies ozone data from a network of Dobson stations over the period 1958-79 and in agreement with several other research group continuously suggested a total ozone decrease between 2.0-4.5 per cent due to nuclear testing effects in the early 1960. But it was controversial. However, it is generally agreed by researcher that future large-scale nuclear exchange would lead to substantial depletion of stratospheric ozone. For, 10,000 million tones nuclear war scenario would be reduced by 30-70 per cent in the northern hemsisphere and by upto 40 per cent in the Southern hemsisphere (Crutzen and Birks, 1982, Turco *et al.*, 1983).

The discovery of stratospheric ozone losses has brought about a remarkably quick international response. At a 1989 conference in Helsinki eighty one nations agreed to phase out CFC production by the end of 20 nm century. As evidence accumulated, showing that losses were larger are more widespread than previously thought the deadline for the elimination of all CFC's was moved upto 1996 and $ 500 million fund was established to assist poor country to switch to non-CFC technologies. Fortunately alternatives to CFC for most uses already exist. The first substitutes will be hydrochlofluorocarbons (HCFC's) which release much less chlorine per molecules. Eventually we hope to develop halogen free molecules that work just as well and are no more expensive than CFC's.

This appears to be an example of quick and effective international cooperation to fight a common environmental threat. Even though researchers disagreed about the scientific evidence on exactly how much ozone loss is due to human activities and what the specific mechanisms of these losses are, the international community got together and agreed an a cause of action.

It is believed that free chlorine released by photolytic dissociation of CFC's reacts with ozone to form molecular oxygen and chlorine monoxide (CIO) and this in turn reacts with atomic oxygen present in the stratosphere to form oxygen molecule and thus regenerate free chlorine. The chlorine may cause damage to human health, animal health. Free chlorine also cause damage to plant health. The increase in concentration of ozone reduced the growth and productivity of plants such as tobacco, tomato, bean, pine and other plants. In California, USA, fruit and vegetable yields have reduced due to O_3 pollution. In the ozone polluted areas, plants are more affected by the insects and diseases.

ACID RAIN OR ACID DEPOSITION

English Scientist Robert Angus Smith coined the term "Acid rain" in his studies of air chemistry in Manchester, England in the 1850. By the 1940 it was known that pollutants including atmospheric acids could be transported long distances by wind. This was thought to be only an academic curiosity until it was shown that precipitation of these acids can have far-reaching ecological effects.

Unpolluted precipitation is commonly assumed to have a pH value higher than 5.6. This mild acidity is caused by the presence of CO_2 in the atmosphere forming carbonic acid. Researchers have suggested that human activities have caused a dramatic increase in the acidity of precipitation at local, regional and a global levels (Likens *et al.*, 1979). Large quantities of oxides of sulphur and nitrogen are being emitted into the atmosphere by the combustion of fossil fuels and industrial processes. These gases are being converted into strong acid (Sulphuric and nitric) which lead to acid precipitation in many areas. Precipitation of lower than 5.6 pH is termed acid rain. Extensive areas of Europe and Eastern northern America are experiencing average pH value of less than 4.5 and even less than 4.0.

Acid rain refers to the process of wet deposition of acidic matter on the earth's surface. In the absence of water, dry deposition occurs. Typical dry deposition rates of pollutants are 0.1 to 1.0 cm/sec for sulphur dioxide, 0.1 cm/sec. far sulphates, 0.2 to 0.5 cm/sec for nitrogen dioxide and 1.0 cm/sec nitrates. Dry deposition may make a major contribution to the acidity problem.

Acid rain is infecting a cocktail of H_2SO_4 and HNO_3. An average 60-70 per cent of the acidity is ascribed H_2SO_4 and 30-40 per cent to HNO_3. The acid rain problem has dramatically increased due to industrialization. Burning of fossil fuels for power generation contributes to almost 60-70 per cent of total SO_2 emitted globally. Emission of NO_2 from anthrogogenic sources ranges between 20-30 million annually over the globe. Acid rain have assumed global ecological problem because oxides travel a long distance and during their journey in atmosphere they may under go physical and chemical transformations to produce more hazardous products.

Many countries like China, India, Romania and Norway have evidence of growth reduction, defoliation, root necrotis, lack of seedling growth and prematuring death. The species affected vary from place to place, but the overall pictures of wide spread plant destruction. Acid deposition thought to leading causes of forest destruction in many areas. Scientists has shown that acid are directly toxic to tender shoots and roots. High altitude forest have intense doses of acids because clouds saturated with pollutants tend to hang on mountain tops, bathing forests in a toxic soup for days or even weeks at a time. Scientists have advise for the other mechanisms may play a role in forest decline. Over fertilization by nitrogen compounds may make trees sensitive to early frost. Essential minerals such as magnesium may be washed out of foliage or soil by acidic precipitation. Toxic metal such as aluminium, may be solublized by acidic ground water. Plant pathogens and insect pests may damage trees or attack trees delibitated by air pollution. Fungi that form essential mutualeistis associations mycorrhizae with tree roots may be damaged by acid rain.

Experiments with simulated acid rain within the observed pH range of acid precipitation have some times decreased crop yield although. Irving (1983) concluded that the effects appear to be very small and when responses were observed they may be positive or negative. Increased incidence of blossom end, roting of tomatoes have been associated with volcanic activity in Hawaii suggesting that acid rain may have brought on a Ca deficiency. Simulated acidic rain on radish plants decreased hypocotyls growth but not shoot growth (Jacobson *et al.*, 1986). Heggostad and Lesser (1990) studied the effects of SO_2 concentration on seedlings of mature soybean. They observed a negative impact on bean yields and seed size. Many bacteria and blue green algal are killed due to acidification, thus distrupting the ecological balance. In West Germany nearly 8 per cent of the forests died and nearly 18 million acres of forests are critically affected by acid rains. Fish population has decreased tremendously and there are deaths of Salman fruit etc. The fishless areas are now fish graveyands.

Acid rain problem is on the avil in India. Industrial areas with the pH value of rain water below or close to the critical value have been recorded in Delhi, Nagpur, Pune, Mumbai and Calcutta. This is due to sulphur dioxide from coal based power plants and petroleum refinery. According to the BARC Air Monitoring Section, the average pH value of acid rain at Calcutta is 5.80, Hyderabad 5.73, Madras 5.85, Delhi 6.21 and Mumbai 4.80. This situation may even worsen in future due to increased installation of Thermal power plants by NTPC. For generating acid problems in India, experiments were conducted in acidic condition on the crop plants shows the treated leaves developed pits, spots and curling. Some times treated plants showed reductions in dry weight as well as seed yield with any symptoms. Seed plant species germinated in acidic atmosphere, the initiation and development of plant disease have shown such as *Cronarium fusiforme* rust of oak, only 14 per cent as many telia formed under acidic rain (3.0 pH). Bean treated with acidic rain (3.2 pH) had only 34 per cent as many nematode egg masses than the acid under a pH 6.0 rain treatment. Bacterial disease (Halo blight) and the rust disease of bean were sometimes more severe and other Milder with the acidic rain.

15

Conclusion

S ince independent India has been making spectacular headway in various spheres and agriculture is no exception to it. As a matter of fact, to a great extent we have succeeded in countering the vagaries of nature. The stigma that the Indian agriculture is a gamble on the monsoon which depicted the helplessness and sorry state of affairs of agriculture in the country no longer haunts us and has been completely wiped out. This has been possible owing to the extensive researches conducted by our agricultural scientists to evolve new methods and techniques in agriculture mainly due to the encouragement and impetus it has been receiving from the government. The priority given to agriculture in the five year plans and the various projects for irrigation and generating electricity in particular covering almost the whole of the country being launched in consequence have minimized if not completely stopped our dependence upon the monsoon. Besides this, new technology has also come to be introduced in the agricultural operations from tilling to harvesting. As a result of this, the country has become almost self-sufficient in the food production. The main problem now is not only to consolidate the present position but also to improve it so as to keep pace with the ever-growing population, a goal that is still eluding, being obstructed and threatened by various natural and man-made factors. The fact of the matter is that the natural resources on account of their senseless exploitation are dwindling. The soil is losing its natural fertility. The crops have become prone to afflictions by pests and diseases. As if to add to the predicament there are natural disasters like floods, drought, storms, cyclones, landslides, earthquakes, volcanic eruptions, wild fire etc. which tend to visit the different regions of the country off and on. Obviously, most of the problems the agriculture sector is facing now have a lot to do with errant and aberrant weather. Therefore, an agrometeorological approach is imperative to sustain the agricultural production.

In fact, we can not come out of the quagmire simply by digging new wells, opening more canals, or constructing fresh dams. It can be done by restraining ourselves from the misuse of our natural resources and keeping the growth of population in check. It is indeed very shocking that the ground water reserve is diminishing at an alarming rate and the water level in most of the regions of the country is going down. The rivers are getting more and more polluted with a large scale factory effluence, tannery wastes, sewage and other toxic substances being released into them regularly, which have rendered the water unfit for drinking and irrigation. The absence of proper mechanism for the conservation of flood water and the wastage of canal water, among others, have but only aggravated the situation. The denudation of landscape caused by indiscriminate felling of trees, cutting of grass and shrub for fuel and fodder as also for raising new colonies, shopping complexes, hotels and holiday resorts, have grossly disturbed the ecosystem particularly the monsoon patter. It has created the condition of drought on one hand while rendering the sea-coast and rivers defenseless against the fury of floods, storms and cyclones, on the other. Not only this, the large scale colonization of the forest area has created tremendous environmental problems as the level of carbon dioxide in the atmosphere has risen beyond proportion giving rise to the phenomena like global warming and carbon dioxide fertilization. The Orissa cyclone (October, 1999) as has been reported in the New Scientist Magazine, assumed the huge and deadly proportion because the mangrove forest which once covered the entire coastline was cleared for economic activities like developing ports, shrimp farms, oil refineries, luxury hotels, and holiday resorts (The Hindustan Times, Dec. 19, 1999).

The above condition does not augur well with the overall health of agriculture in the country is not gainsaid. As if to worsen the situation further, there is the menace of other natural calamities like fog and frost, pests and diseases, so on and so forth. To save the agriculture against the threats posed by the nature as such two-fold strategies, preventive and remedial, are warranted. The preventive measures calls for adopting such actions beforehand as to see that the ecological balance is properly maintained so as the inclement weather conditions referred to above do not occur or at least their intensity is kept under control. On the contrary, the remedial measures are in fact a combating mechanism to deal with a particular natural calamity as and when it arises so as to minimize the losses both in men and material.

Curiously, both preventive and remedial measures are directed by one single major factor *viz.* the mood of the weather. To be more articulate, if we know about the pattern in which the weather is going to behave in a foreseeable future, it becomes possible to plan our activities regarding various agricultural operations accordingly. In view of this, the science and mechanism of weather forecast are becoming increasingly important and large-scale researches are being conducted by agricultural and environmental scientists on its various aspects which come in the scope and purview of agrometeorology.

In this backdrop we have endeavoured to understand as to how an agrometeorological approach to agriculture can help in various ways to achieve a sustainable and continued growth in the food production without disturbing the ecological balance. To make this a reality, the need is of conserving the natural resources and the purity level of the environment. Because it is not possible to achieve the goal of boosting the food production if we continue to indulge in the indiscriminate exploitation of the natural resources like clearing the forest for short-term economic gains, wastage of water resources particularly underground, river and rain water, over-cropping of the land without caring for the maintenance of its fertility level etc. Besides, the environmental pollution is another major impediment in the food production. Large scale carbon dioxide being released in the atmosphere in the form of chimney and motor vehicle smoke has brought about unusual increase in the level of carbon dioxide known as the carbon dioxide fertilization which is caused by the disequilibrium in its inhaling and exhaling process. As if to deteriorate the situation further, there is the water pollution caused in various ways.

In fact the relationship between man and nature like any human relationship is based on reciprocity. The only difference, however, is that the nature bestows its bounties upon man without demanding anything in return. Truly speaking, man has nothing to offer to nature and he is the sole and one sided beneficiary in the above relationship. The maximum the nature requires is that the man should not cross the limit while availing the gifts of nature. That is to say, he should enjoy the benefits of nature only to the extent that it does not become extortion. If it is not so, the ecological balance is disturbed which not only casts adverse effects on the agricultural production but brings about such serious consequences as flood, drought, cyclone etc. which tend to put the very human existence in jeopardy.

The agrometeorological approach to agriculture in the first place makes it incumbent to keep in mind the regional or local weather conditions and geographical peculiarities. As a matter of fact we cannot ignore them while planning our agricultural operations including the selection of crop to be grown. The agricultural operations vary from region to region and so are the inputs like irrigation, fertilizers, cultivators and other such equipments. To put it more candidly, in the river bed area the soil being naturally soft, wet and fertile, the cost of tilling, irrigation and fertilization is minimal as compared to upland area. Similarly, the weather conditions like the duration and time of occurrence of the monsoon and its quantum as also the physical conditions of the particular region per se decide as to which crop and variety should be grown in that region to obtain optimum returns. It is in view of this that India has been divided into several zones. Dr. M.S. Randhawa, and eminent personality, a bureaucrat turned scholar and agricultural scientist, has sought to place India into five agricultural zones on the basis of climate, crops and stock animals which help plan the agriculture activities accordingly. The current weather conditions and the likely change in the climate also results in the proportional changes in the production. By way of crop simulation models we can understand and anticipate the position of production beforehand which naturally enable to adopt remedial strategies in the event of prospective loss in the production. In this way the proper utilization of local resources and the adoption of strategies commensurating with the climatic change which comes within the scope and purview of agrometeorology can be of great help to save the production from going down.

The meteorological information also helps a lot in fighting against the danger of pests and diseases to the crop. Since it can be precisely understood both by experience and experiments as to which type of weather conditions are conducive to the outbreak of the particular disease or pest on the basis of prospective change in the climate a definite idea can be formed about them well in advance to evolve proper strategies to counter them effectively. This we have tried to illustrate by was of various examples of diseases likely to afflict a particular crop in particular weather condition which clearly proves that the weather information can help the farmer to save the crop from pests and diseases.

The meteorological information's can also be very helpful at macro level. The problem of ecological imbalances has assumed a gigantic and horrifying posture. The spread of materialistic culture manifesting in the form of over-industrialization, colonization and unrelenting urge of man to tame and subdue nature, has created more problems for man than solving them. The root cause of this has been the phenomenal growth in the world population particularly in the latter half of the 20th century. In certain parts of the world in which Indian sub-continent and China figure at the top, the spurt in the population has been relatively very high. As a result of this the strain on the natural resources in these regions has been tremendous. In view of this, the prime need now is to conserve the natural resources which are possible only when they are used properly and sensibly. One way of doing this is to avoid the vagaries like deforestation, water and air pollution and also leading an artificial life *i.e.* a life which is not compatible with the natural conditions. In other words, our life pattern should be eco-friendly and our guiding principle should be that what we take from nature, so to say that what nature affords to us, must be repaid. The attitude that how much and how best we can extract from the nature without doing anything in reciprocity must be avoided.

Hopefully, now the need for this is being realized world over but not before paying a heavy price for that. For, due to the increase in the CO_2 level, the respiration system has been totally impaired which has brought about the phenomenon called as the global warming. This, among others, is likely to cause serious adverse effects on agricultural production. Therefore, it is necessary to assess the impact of this global climate change on the crops to enable us to evolve counter-strategies to deal with them, which include the developing of new crop species and agricultural methods and parameters for the preparation of soil, fertilization, irrigation etc. to cope with them. For this purpose, a concerted and co-ordinated attempt on the part of the scientists of the world is essential. It is obviously in pursuance to the same that the institutions like Inter-governmental Panel on Climate Change (IPCC) have been founded to take stock of the global climate change so as to find out, among others, the alternative crop species and combination for different regions. As mentioned earlier, it is in one of the reports of the IPCC that such starting revelation has been made that the forests cannot be treated as permanent carbon sinks and that they cannot be taken for granted to solve the problem of global warming. This sounded a note of caution among scientists and environmentalists the world over the devise alternative strategies against the CO_2 elevation so as to mitigate its ill effects. Besides, this there is another mechanism known as Expert Decision Support System (EDDS) whose job it is to check the factors which hold the key to climatic change affecting biosphere adversely so that the crops can be saved.

The weather forecasts being so much important for planning the agricultural operations at each and every stage, the need was felt to evolve proper system and mechanism for the same. In view of this, meteorological observatories have been established both at national and international levels with its units at the regional or district level. For an accurate forecasting, it is expedient to send the local weather data to the National Data Collection Centre (NDCC) for confirmation duly comparing and processing it. The weather forecasts, in view of their periodical relevance, can be of three terms-short, medium and long. The short term forecasts are helpful in planning the instant agriculture activities like the tillage, irrigation, fertilization and spraying of medicines. Therefore, they are also known as operational forecasts. The medium forecasts called also as the planner forecasts are helpful in planning the agricultural operations valid for more than a week say for 10 days or so, while the long range forecasts cover a period ranging from above 10 days to a month. Sometimes the long range forecasts can include a period of that of a full season. This forecast is mainly based upon the data record of the past which enables the farmers to select a particular variety of seeds and make preventive strategies to fight the impending towards weather conditions.

Moreover, in the event of probable meteorological disasters like flood, drought, storm, cyclone, landslide, earthquake etc. the weather forecasts can help a lot in fighting and managing them by way of alternative strategies. With the advent of advanced technologies like computers, satellite, radar etc., it has become rather easier to take effective defense measure against these natural calamities. The computer technology in particular is very important in this respect as it preserves various data which help in advance planning and decision making. The development of efficient computer system has given a new dimension to the planning and disaster management process. The Geographical Information System (GIS) which is a computer based system for storing, editing, displaying and plotting geographical data can help in creating Land Information System (LIS), Water Information System (WIS), and Disaster Management Information System (DMIS) thereby giving information's about various resources like land, water, vegetation and socio-economic conditions.

The weather forecasts can thus help in fighting the natural disasters like cyclone and earthquake, and flood, drought, fog and frost, hot and cold waves, dust and hail storms etc. It also helps the government to make appropriate measures for various relief measures to be initiated in the given situations in this ways the weather forecasts are equally important for farmers, policy makers as well as relief workers at various level.

Apart from the weather forecasts there are Agrometeorological Advisory Services (AAS) for advising, educating, training and guiding the farmers. This is, in fact, an inter-disciplinary activity which includes all the departments of agriculture such as agronomy, soil science, plant pathology, entomology, horticulture, and meteorology. All these departments conduct co-ordinated researches to develop crop varieties and cropping system in a given area keeping in views its physiology and weather conditions. The AAS hold workshops and training programmes for the farmers to create greater awareness among them as to which method and crop species are favourable to their area. It also forewarns them about the probable attack of pests and diseases. Not only this, the periodical bulletins known as Agrometeorological Bulletins (AAB) are also issued to help the farmers to plan their agricultural activities accordingly. It is needless to say that these bulletins serve as a ready reckoner for the farmers to help them measure their each and every step and find out which direction should they proceed in, to elicit optimum returns in the given conditions. It is thus obvious that a multi-disciplinary approach to solve the problem facing the agriculture and its process in a region or area in very essential and so a coordinated effort in the field of research and development. This will enable the farmers to get acquainted with the advance methods and techniques of agriculture to help them achieve the goal of sustainable agricultural production.

In a nutshell, weather plays the most dominant role, far more crucial and decisive than that of man, in shaping the form of various economic activities, more so the agriculture. But the credit goes to the wide researches conducted world over on how to obtain optimum benefits form the natural resources without destroying the ecosystem or ecological balance. The crux of the problem, however, is that our natural resources are limited, whereas our population is increasing by leaps and bounds. Thus, to provide sustenance to our teeming millions making them available food, cloth and shelter has been a major challenge for both the rulers of the time as well as the leaders of the society having a very glorious past, we cannot sit contented merely by availing minimum required level of living. After all we have also to keep pace with all the developed countries of the world. We have tremendous potential in human resources both quantity and quality wise. We possess and command a high level of expertise in various fields. Not only our intellectuals and philosopher but also engineers, doctors and technocrats have made their presence felt in the world arena. But this is only the irony of fate that we still lag far behind in per capita income than even many smaller courtiers of the world. The reason for this paradox is not difficult to find. Our economy is based mainly on agriculture as more than 70 per cent of the population accounts of its dependence upon this sector. As pointed out earlier, the root cause of all our predicament is the fast growth of our population. As a result of this, the burden on our natural resources has been continuously increasing. But in a bid to bring new and larger area under cultivation, when the forests were begun to be cleared, it brought about serious climatic changes. For instance, due to this the monsoon pattern stood changed, the level of carbon dioxide in the atmosphere began to increase, the soil became increasingly prone to erosion, and some of the animal species which played a pivotal role in striking a balance in the ecosystem began facing extinction. The land began constantly losing its natural fertility and the overdose of chemical fertilizers to tone up the same began leaving toxic influences. It is as a result of this that the food grains are becoming unhygienic and even milk and milk products are losing their natural taste and flavour. This process is still continuing and so are its ill-effects thereby turning the situation from bad to worse. In this way, the land is being unduly exploited and the level of fertility is being maintained through artificial means which has cast serious ill-effects on the environment in ways more than one. But, on the contrary, in many parts of the country vast tract of land is still lying barren. Under these circumstances it is not unduly felt that such areas should be brought under cultivation by making proper arrangements for its leveling, irrigating etc. rather than felling trees and clearing forests. For, the indiscriminate felling of trees is going to create such serious problems as flood, drought, storm, cyclone, and landslide etc. on one hand, while on the other hand it is likely to cause large-scale pollution giving rise to serious health problems and the global warming which tend to disturb the entire natural balance. As pointed out earlier, the new plantations are not going to solve the problem of carbon dioxide fertilization fully. In fact, before industrialization natural forces were in equilibrium with the atmosphere but the idea of artificial forestation, though extremely useful under the circumstances obtaining as of now, cannot be a substitute to the old forests. It is a matter of great concern that the damage which has already been done by deforestation cannot be compensated by any human efforts whatever. This, however, does not mean that we should give up the idea of raising forests because it will be useful if not so much to the present generations then of course for the generations to come. What we want to emphasize is that while raising new forests we ought to stop destroying the old ones. The idea of wiping out the old forests, even if under the surety of replacing them by way of planned forestation, is dangerous. It is not wise to invite large-scale devastation for petty gains as has been the case in Orissa. The study has revealed that the recent catastrophe that overtook Orissa in the October last was a man-made problem. It is thus observed that "The havoc the October super-cyclone caused in Orissa could have been avoided, had its mangrove forests not been destroyed to develop shrimp farms". (The Hindustan Times, Dec. 19, 1999).

In this way while augmenting our efforts to boost agriculture production what Dr. Swaminathan has preferred to call as bringing about the Second Green Revolution, it has to be borne in mind that this should not be done on the cost of our natural assets and the environment. Therefore, what is needed is that we should strive to increase agriculture production without playing with our environment which has already suffered a lot having fallen victim to human need, nay greed. For this, we are required to be more and more eco-friendly and to give up the idea of taming and enslaving the nature in a calculated manner unmindful of its serious repercussions. In this regard we should always be guided by meteorological tips and briefs, and remain prepared to take steps keeping in view the mood of the nature. This does not mean that we should avoid using modern means and techniques thereby reversing the process of modernization but this should be done only to the extent of that it does not come in the way of the ecological balance and natural harmony. This is the only safer way to achieve the goal of sustainable agriculture as also for maintaining the pace of food production to meet our ever-growing requirements.

References

Abel, P., Nelson, P, De R.S., Hoffman, B., Rogers N, Fraley, S.G. and Beachy, R.N. (1986). Delay of disease development in transgenic plants that express the tobacco mosaic virus coat protein gene. *Science* 232: 738-743.

Abelson, P.H. and Hammond, A.L. (1974). Energy use conservation and supply's *Science*.

Ahuja, V. and Tiwari, R. (2003). Biosorption of Pb and Cd by free and Immobilized Biomass of Acinetobacter anitratus. Role of microbes in the management of ewnvironmental pollution eds. R. Tiwari, K.G. Mukherji, J.K. Gupta, L.K. Gupta. A.P.H. Publishing Corporation, New Delhi, India, pp. 149-156.

Ali, S. and Watanable, I. (1986). Response of Azolla to P, K and Zn in different wetland rice soils in relation to Chemistry of flood water. *Soil Sci. Pl. Nutr.*, 32: 239-253.

Babich, H. Davis, D.L. and Stotzky, G. (1980). Acid precipitation : Causes and Consequences. *Environment* 2, 6-12, 40.

Bach, W. (1978). The potential consequences of increasing carbon dioxide levels in the atmosphere. In J. Williams (8), Carbon dioxide, cimate and Society Oxford: Pergamon Press, 141-67.

Bagyaraj, D.J. (1990). Biological interaction between VAS fungi and other beneficial organisms. In current trend in mycorrhizal research, Proceedings of National Conference on mycirrhiza held at H.A.U., Hisar, Feb. 14-16, 1990 pp. 76-77.

Belliveau, B.H.;Starodub, M.E. and Trevors, J.T. (1991). Occurrence of antibiotic and metal resistance and piasmide in *Bracillus* strains isolated from marine sediment. *Con. J. Microbial*, 37: 513-520.

Bethlen Ferlvay, G.J.; Brown, M.S.; Ames, R.N. and Thamas, R.S. (1988). Effect of drought on host and endophyte development in mycorrhizal soybean in relation to water use and phosphate uptake. *Physiol Plant*, 72: 565-571.

Brietey, C.L.; Killey, D.P.; Seal, K.J. and Best, D.J. (1985). Materials and biotechnology, Principles and Applications eds. I.J. Higgins, D.J. Best and J. Jones. Blackwell, Oxford, pp. 163.

Carling, D.R.; Riehle, W.G.; Brown, M.; Johnson, D.R. (1978). Effect of vasicular arbuscular mycorrhizal fungus on nitrate reductage and nitrogenase activities in nodulating and non-nodulating legumes. *Phytopathology*, 68: 1590-1596.

Changnon, S.A., Huff, F.A. Schickedaz, P.J. and Vogel, J.L. (1977). Summary of METROMEX Volume 1. Weather on malies and impacts. Urbana. Illinois State, water survey division Bulletin No. 62.

Crutzen, P.J. and Birks, J.W. (1982). The atmospheric after a nuclear war: twilisht at noon: *Ambio*, 11: 114-25.

Cuozzo, M., Keith, M.O. Connell, Wojciech, K, Rong, F.F., Mana, H.C. and Nilgun, E,T. (1988) Viral protection in transgenic tobacco plants expressing the cucumber mosai virus coat protein or its antisens RNA, *Biotechnology* 6: 549-557.

D, Amore, T. and Stewart, G. C. (1987). *Enzyme Microb. Technal.* 9 : 322-326.

Edwards, C.A. (1973). Persistent pesticides in the environment (R.C. Prezd. Cleveland Znd End.

Einbencler, G. Bakaliors, A. Wall, T. Hoagland, P and Kamlet, K.S. (1982). The case on immediate controls on acid rain, materials and society, 6, 251-282.

Elsom, D. (1987). Atmospheric pollution. Basil Black Well Ltd.

Fillari, J.J., Kisore, J., Rose, R., and Comai, L. (1987). Efficient transfer of a glyphosate tolerance gene into tomato using a binary agro bacterium tumefactions vector. *Biotechnology* 5: 726-730.

Fischhoff, F.A., Bowdish, K.S. Perlak, F.J., Marrone, P.G., Mccormick, S.M., Niedermeyer, J.G. Dean, D.A., Kausano, K, Kretzmer, Mafer, E.J. Rochester, D.E., Rogers, S.C. and Fraley, R.T. (1987). Insect tolerant transgenic tomato planks. *Biotechnology* 5: 807-813.

Fischoff, D.A. (1987) Biotechnolgy 5: 807-813.

Gadd, G.M. (1992). In microbial control of pollution eds. J.C. Fry, G.M. Cadd, C.W. Jones, R.A. Stobert and I. Watso-Craik Cambridge Univ. Press, Cambridge, pp. 59-88.

Garg, N.; Garg, K.L. and Mukarji, K.G. (1994). Microbial degradation of envieronmental pollution. Recent Advances in Biodeterioration and Biodegradation eds. K.L. Garg, Neelima Garg and K.G.Mukarji. Nayer Prakashan, Culcutta, India.

Gilbert, O.L. (1974). An air pollution survey by school children. *Environ. Pollution*. 6, 170-180.

Gorham, E. (1982). What to do about acid rain tech. *Review*, 82, 59-70.

Gupta, R. and Mukherji, K.G. (2003). Bioremediation; past, present and future, Role of microbes in the management of environmental pollution eds. R. Tiwari, K.G. Mukherji, J.K. Gupta and L.K. Gupta. A.P.H. Publishing Corporation, New Delhi-02, India.

Handow, S.E., Panoponlos, N.S. and Mefarland, B.R. (1989). Genetic engineering of bacteria from managed and natural habitats. *Science* 244: 1300-1307.

Hansen, J.; Johnson, D.; Lacis, A.; Lebedeff, S.; Lee, P.; Rind, D. and Rusell, G. (1983). Climatic effects of atmosphere carbon dioxide, *Science*, 220, 873-5.

Hawksworth, D.L. and Rose, F. (1976). Lichens as pollution monitors. London: Edward Arnold, Inst. of Biology, Studies in Biology No. 66.

Hayes, W.A. (1968). Microbiological changes in composting wheat straw/horse manure mixtures. Mushroom Science, VII, 173-186.

Helden, V., Gatehouse, A. Sheerman, S. Baker, R. and Boulter, D. (1987). A novel mechanisms do insect resistance engineered into tobacco. *Nature* 330: 160-163.

Hemenway, G., Fang, R.X. Kante Waski, W.K. Chua, N.H. and Tumer, N.E. (1988). Analysis of mechanism of protection in transgenic plants expressing the potato virus x coat protein or its antisense RNA, EMBO. *Journal* 7: 1051-1059.

Hinrichsen, D. (1986). Multiple pollutants and forest decline. *Ambio.* 15, 258-65.

Hulm, P. (1982). WHO reiterates risk of ozone depletion. *Ambio*-11-70.

Kantikar, T. and Mistry, M. (2000). Status of women in India, In interstate comparison. *Indian Journal of Social Work* 61: 366-383.

Leach, J.E. and White. F.F. (1989). Analsysis of xanthomonas compesnis pv. Oryzal role paper presented during the third ann. Mtg. of the Rockefellor Foundation International Programme on Rice Biotechnology St. Lovis Missouri USA.

Leung, H.R. Nelson, a., Amante, N., Oliva, E., Borromeo, A.K., Roymundo, E., Ardales, R.D., Dalmcio and Sitch, L.A. (1989). Molecular rice pathology research at IRRI, paper presented during the 3rd ann. Mtg. of the Rockefellor Foundation International Programme on Rice Biotechnology St. Lovis Missouri USA.

Lie, S.C.; Cicerone, R.J.; Donahue, T.M. and Chamides, W.L. (1976). Limitation of fertilizer induced ozone reduction by the long life-time of the reservoir of fixed nitrogen. *Geophy, Res. Letters*, 3: 157-60.

Likens, G.E.; Wright, R.F.; Galloway, J.N. and Bunga (1979). Acid Rain. Scientific American, 242: 39-47.

Lumpkin, T.A. and Plucknett, D.L. (1982). Azolla as a manure use and management in crop production. West View Press Colorado, U.S.A.

Madhyastha, M.N. (2003). Prospects and problems of environment, Daya Publishing House, Delhi, India.

Mallick, S. and Kushari, D.P. (1990). Effect of different potassium concentrations on the growth and bioaccumulation of elements of different *Azolla* species. *Proc. 3rd Ind. Sci Cong.*, 32-33.

McElroy, M.B.; Elkins, J.W.; Wohsy, S.C.; Yung, Y.L. (1976). Sources and sinks for atmospheric nitrous oxide. *Rev. Geophys. Spoce Phys.*, 14: 143-50.

Meargeay, M. (1991). Towards an understanding of the genetics of bacterial metal resistance. *TIBTECH*, 9: 17-24.

Miller, P.R., Parmoter, J.R., Flick, B.H. and Martinez C.W. (1969) Ozone damage response of ponderosa pine seedlings. *J. Arr. Pollut. Control Ass.* 19, 435-8.

Mishra, C.M.; Srivastava, R.; Dubey, P. and Burpal, B.S. (1999). Role of Tree species in pollution abalement. Advances in Environmental biopollution (eds.) A.C. Shukla, Bandana, A., Trivedi, P.S. and Pandey, S.N. A.P.H. Publishing Corporation, Darya Ganj, New Delhi, India, p. 357-361.

Mishra, R.S. (1992). Comparative studies of Sewage and tubewell irrigation on the rhizosphere and non-rhizosphere microflora special reference to brinjal and chilli crop. M. Sc. (Ag.) Thesis, A.A.I., Allahabad.

Mishra, R.S. Shukla, P.K. and Singh, R.V.C. (2007). Mycoflorce of compost and casing soil of white button mushroom. *J. Mycol. Pathol. Vol. 37* (3): 424-425.

Morgan-Huws, D.I. and Haynes, F.N. (1973). Distribution of same epiphytic lichens around an oil refinery at Fawlew, Hampshire. In B.W. Ferry, M.S. Baddley and D.L. Hawks worth (eds.) Air pollution Lichens. London: Alhlone Press 89-108.

NORC (1979) carbon dioxide and climate a scientific assessment Washington, DC National Academic Press, Ad Hoc Study Group.

Nyamaptene, K.W. and Mtetwa, W.G. (1987). Water, air and Soil pollution 27: 401-411.

Okon, Y. (1985). Azospirillum as a potential inoculant for Agriculture. *Trends in Biotech*, 3: 223-228.

Panwar, J.D.S. (1991). Effect of VAM and Azospirillum brasilens on photosynthesis, nitrogen metabolism and grain yield in wheat. Indian *J. Plant Physiol.*, 34 (4): 345-361.

Panwar, J.D.S. and Sirohi, G.S. (1989). Azospirillum-An important biofertilizer for improving crop productivity. *Farmers and Parliament*, 24: 23-24.

Panwar, J.D.S.; Pandey, M. and Abrol, Y.P. (1990). Effect of Azospirillum brasilens on photosynthesis, transpiration and yield of wheat under low fertility conditions. *Indian J. Plant Physiol.*, 33: 185-189.

Paramafivan, T. (2002) taking protissional care of the nature, the environmental audit. *J. Institute of charted Accountants of India* 51, (2) 148-155.

Pushpa Srivastava (1999). Algae in abatement of pollution II. Textile factory effluent. Advanced in environmental bio-pollution. A.C. Shukla, Vandana, A., P.S., Trivedi and S.N. Pandey. A.P.H. Publishing Corporation, New Delhi, India. Page 65-70.

Rai, V.N. (1992). Toxicity and bioaccumulation of chromium in a chlorococcalean green algae. Gilaucolystes nastichineorum I Asia Pacific Conference on Algae Biotechnology 29-31 Jan. 92, Kualalumpur.

Ramos, J.L. Wasserfallen, A. Rosi, K. and Timmis, K.N. (1987). *Science* 235: 593-596.

Rana, B.C. (1995). Studies on chemical and biological treatment of a zinc smiltten effluent its evaluation through the growth of test algae Nava Hedwigra 23: 465-471.

Rani, H.L. and Lalitha Kumari (1994). Degradation of methyl parathion by Pseudomonas putida. *Can. J. Microbial*, 4: 1000.

Record, F.A., Budenick, D.V. and Kindya, R.J. (1982). Acid rain information Book New Jersey Noyes Data Corporation.

Reinsel, G.C. (1981). Analysis of total ozone data for the detection of recent trends and the effects of nuclear testing during the 1960s. *Geophys. Rest. Letter*, 8: 1127-30.

Revards, Y. and Cailleux, R. (1972). Contribution ai-etude des micro organisms du compost destine ala culture du champignon de couche. *Reve de Mycologie* 37: 36-37.

Revelle, R.r. and Shapero, D.C. (1978). Energy and Climato, Environ. Conservation, 5, 81-91.

S Korby, L and Selden, G. (1984). The effect of ozone on crops and forests Ambio. 13, 68-72.

Sahu, J.K.; Nayak, H. and Adhikary, S.P. (1996). Blue green algae of rice fields of Orissa state. Distributional pattern in different agro-climatic zone. *Phykos*, 35: 67-84.

Sandhach, F. (1982). The pollution problem. In: principles of pollution control, Long Man Group Ltd. pp. 1-21.

Shah, D.M., Horseh, R.B., Kleu, H.J., Kishore, G.M., winter, J.A., Tumer, N.E. Horonaka, C.M., Sandoz, P.R., Gasser, C.S., Aykent, S, Siegd, M.R., Rogers, S.G. and Fraley, R.J. (1986). Engineered herbicide tolerance in transgenic plants. *Science*, 233: 478-481.

Sharma, M. and Srivastava, P. (1994). Influence of Saras dairy effluent on the growth of Spirulina Subsalaoert Ex Gromont. *J. Environ. Poll.* 1 (2): 55-59.

Shaver, C.L., Cass, G.R. and Druzik, J. (1983). Ozone and the deterioration of works of art. *Environ. Sci. Technology*, 17, 748-52.

Silver, S. (1981). Environmental and speciation and monitoring needs for trace metal containing substances from energy related processes eds. F.E. Brinkman and R.H. Fish. NBS Publications, pp. 301-311.

Sionit, N, Strain, B.R. Hellmers, H and Kramer, P.J. (1981). Effect of atmospheric carbon dioxide concentration and water stress on water relations of wheat. *Botanical Gazette*, 142, 191-196.

Stratsoma, G.; Samson, R.A.; Olijusma, T.W.; OpDencamp, H.J.N. and Van Griensvan, L.J.L.D. (1994a). Ecology of thermophillic fungi in mushroom com[post with examphasis on S. thermophillium and grow stimulation of *A. bisporus* mycelium. *Appl. Environ. Microbial*, 60: 454-458.

Temple, P.J. and Taylor, O.C. (1983). World vide ambient measurements of peroxy a cetyle nitrate (PAM) and implelications for plant injury. *Atmosphere Environ.* 17, 1583-7.

Tumer, N.E., O. Connell, K.M., Ne,son, R.s. Sanders P.R., Beachy, R.N. Fraley R.T. and Shah, D.M. (1987). Expression of altalata mosaic virus coat protein gene confers cross protection in transgenic tobacco and tomato plants. *EMBO. J.* 6: 1181-1188.

Turco, R.P.; Toon, O.B.; Ackerman, T.P.; Pollack, J.B. and Sagan, C. (1983). Nuclear winter: global consequence of multiple nuclear explosions, *Science* 222, 285-93.

Ulrich, B. and Pankraklh, J. (1983). Effect of accumulation of air pollutants in forests eco-systems. *Bosten Reidel.*

Ulrich, B. Mayer, R. and Khanna, P.K. (1980). Chemical changes due to acid precipitation in a loss derived soil in central Europe. *Soil Science* 30, 193-9.

United Nations Economic Commission for Europe (1985). Air Pollution across Nations Boundaries New York: United Nations, Air Pollution studes-2.

V.K. Royal Commission and Environmental Pollution (1971) First Report: Pollution: HMSO.

Vaeck, M., Reynaerts, A., Hofte, A., Jansens, S., Baukeleer, M.D., Dean, C., Zabeay, M., Montagu, M.V. and Leemans, J. (1987). Transgenic plants protected from insect attack. *Nature* 328: 33-37.

Vaughan, A. Angharad, M.R., Gatehouse, Suzanne, E., Sheerman, Richard, F. Barkers and Donald Boulter (1987). N Novel mechanism of insect resistance enginnered into tobacco. *Nature.* Vol. 300 (12): 140-143.

Vijay, B. (1996). Investigation compost mycoflora and Crop Improvement in Aquaricus bisporus (Longe). Singh Ph. D. Thesis, Himachal Pradesh University, Shimla, India.

Waggoner, P.F. (1984). Agriculture and carbon dioxide Amer. *Scientist* 72, 179-84.

Wark, K. and Warner, C.F. (1981). Air pollution its orizins and control, 2nd edition, New York : Harper of Row.

Wase,A.J. and Forster, C.F. (1997). Biosorbent for metal ions, Taylor and Francis, London.

Watanable, I. and Ramireza, C. (1984). Relationship between soil phosphorus availability and Azolla growth. *Soil Sci. Plant Nutr.,* 30: 595-598.

Weber, E. (1982). Air pollution: assessment methodology and modeling vol. 2 New York: Plenum Press.

Wetstone, G.S. and Foster, S.A. (1983). Acid precipitation: What is it doing to our forests. *Environment,* 25, 10-12, 38-40.

Whitton, A.B. (1994). Algae as monitors of water pollution 2nd Asia Pacific Conference on algae biotechnology, 25-27 April, 1994. National University of Singapur, 47.

Wolfe, M.S. and Barrett, J.A. (1980). *Plant disease* 64: 148-155.

World Health Organization (1976b). WMO stolement on modification of the ozone layer due to human activities. *Environ. Conservation,* 3, 68-70.

Index